AIR LOGIC CONTROL
for
AUTOMATED SYSTEMS

T0203751

Rudy Wojtecki, P.E.

Assistant Professor
Mechanical Engineering Technology
School of Technology
Kent State University

CRC Press
Taylor & Francis Group
Boca Raton London New York

CRC Press is an imprint of the
Taylor & Francis Group, an **informa** business

CRC Press
Taylor & Francis Group
6000 Broken Sound Parkway NW, Suite 300
Boca Raton, FL 33487-2742

First issued in paperback 2019

ISBN-13: 978-0-8493-2057-6 (hbk)
ISBN-13: 978-0-367-39965-8 (pbk)

Library of Congress Cataloging-in-Publication Data

Wojtecki, Rudy G.
    Air logic control for automated systems / Rudy G. Wojtecki.
        p.   cm.
    Includes bibliographical references and index.
    ISBN 0-8493-2057-7 (alk. paper)
    1. Pneumatic control.   2. Logic circuits.   I. Title.
TJ219.W64   1999
629.8'045--dc21   99-25433
                                                                                      CIP

International Standard Book Number 0-8493-2057-7
Library of Congress Card Number 99-25433

Visit the Taylor & Francis Web site at
http://www.taylorandfrancis.com

and the CRC Press Web site at
http://www.crcpress.com

Dedication

This book is dedicated to my father, Rudy Wojtecki Sr., who taught me that the secret to success is hard work.

Preface

Air Logic Control (ALC) is used in a variety of industrial manufacturing operations and is especially useful for automated systems which utilize fluid power devices and can easily be interfaced with PLCs and other electronic controls. As industrial processes become more and more automated, the use of ALC appears to be increasing. With the ever increasing amount of ALC in use, the need exists for people to have a working knowledge of this technology.

This book is a result of the need for a text to provide appropriate and contemporary material for a special topics course in ALC offered by Kent State University. The discussions presented are intended to serve as a fundamental text for this course. The material presented is designed for the technology student and assumes the student has a background in physics and basic mathematics as well as fluid power. Even though the material is targeted for a university level curricula, the discussions are presented in as practical a manner as possible to allow the book to be used as a design reference as well.

The objective of the text is to provide the necessary background information which will allow the student to build a solid understanding of Air Logic as well as discuss contemporary concepts used in industrial control systems. General areas of discussion include theory, components, and applications. Even though some discussion focuses on design, the intention of these discussions is not to limit the usefulness of the text to learning the design process but to apply design techniques to understanding ALC from the point of view of implementation, programming, maintenance, and troubleshooting.

Three case studies are discussed in progressing complexity. These studies contain redundant approaches to the problem solution to provide the student with alternative methods of solution. In practice it is a personal preference based on experience as to which method works best.

I would like to thank the people who have supported my effort in writing this book and especially thank the General Motors Apprentices who endured my presentations of the material in this book before it was even written. Their help and support has been of great value in making this text a reality.

Rudy G. Wojtecki

Table of Contents

AIR LOGIC CONTROL
for
AUTOMATED SYSTEMS

1 Introduction to Air Logic Control

Fierce competition requires manufacturers to constantly improve their product quality and at the same time improve their production efficiencies by reducing manufacturing costs and shortening cycle times. This competitive pressure is responsible for the increasing implementation of production automation. The development of powerful, inexpensive computers and Programmable Logic Controllers (PLCs), dedicated computers used for shop floor control applications, has allowed the development of extremely complex and sophisticated automated systems. This need for automation fuels the need for control systems which enable processes to perform their intended functions. These control systems, to function satisfactorily in the manufacturing environment, must be reliable, rugged, and inexpensive. It is common to associate control systems with electrical/electronic systems. Even though electronic logic systems dominate industrial controls, many applications can be simplified and operated more effectively using air logic control (ALC) especially pneumatic system control where the energy source for control can be common with the power source. In many shop floor applications, ALC is more reliable than electronic control. The function of ALC is to gather and process information and to provide pressure signals which can manipulate power devices to do the work they are intended to do when they are to do it. Air is a convenient fluid for energy conversion and transmission. The availability of air is unlimited and the cost of this fluid is inexpensive. Using air in pneumatic systems is nonpolluting and does not harm the atmosphere.

1.1 HISTORY

Air pressure (wind) has been used to power ships since the beginning of civilization. Bellows were used to produce bursts of compressed air to start fires as early as 3000 B.C. In the 1700s musical instruments were controlled by mechanical drums which operated valves to switch in pipes of different pitches. This concept evolved in the 1800s into pneumatic controls using perforated cardboard strips and by attaching long strips end to end continuous control was developed which could play the programmed melody over and over. Also the strips were eventually replaced by long paper rolls which could play lengthy and complex combinations of songs. The control system was placed on pianos and we know them as "player pianos" (Figure 1.1). This technology was adapted to automate the textile industry in the early nineteenth century. In the early 1800s several railroads were built in Europe which operated by pneumatic power and pulled loads as fast as 42 mph. By 1858 compressed air drills were being used in mining. A famous application of pneumatic control was developed by Westinghouse in the late 1800s to control the stopping of railroad trains. Air brakes are used on most large vehicles today including trucks.

1

Compressed air was used by many young boys in the late 1800s for entertainment to fire corks from toy rifles called "pop-guns." Much of the development in steam systems was a result of earlier work in pneumatics. The success of steam power was due to the invention of the lathe (Wilkensen's boring machine) which allowed cylinders with concentric bores to be made. The ability of industry to make good quality cylinders allowed progress to be made in the development of pneumatic equipment as well as the steam engine.

Use of compressed air preceded the use of steam and electricity as a control medium. In the 18th and 19th centuries a great deal of development took place utilizing air as a power source and also as a means of control. Pneumatics was a primary power source in many cities in the late 1800s. Paris had over 30 miles of pipe conveying 90 psi air throughout the city. This air system was supplied by a 25,000 HP compressor. At that time the use of electricity was emerging but technical problems limited its use. Late in the 19th century the development of electrical equipment upstaged the use of air power and control and the use of pneumatics subsided.

The basis for much fluid logic technology was developed in the mid 1950s at Oklahoma State University. Pneumatic systems were used primarily as power sources until early 1960s when fluidics devices were developed and introduced to the control market. This technology was oversold and overrated and many applications problems soon developed. Because of misunderstanding and problems due to misapplication, engineers lost confidence in the use of air as a control means. Until the recent emphasis on automation air control was not of interest to most engineers. In the last few years a large number of control devices have been developed and commercialized which provide the equipment designer with an interesting selection of devices.

1.2   INDUSTRIAL APPLICATIONS

The equipment designer has several choices for types of control systems:

- manual
- mechanical
- electrical
- hydraulic
- pneumatic (ALC)
- combinations

Each type has advantages and disadvantages. The ultimate selection criteria are the requirements of the application. Application requirements can include control:

- accuracy
- resolution
- repeatability
- reliability
- response time
- safety both to personnel and other equipment
- life

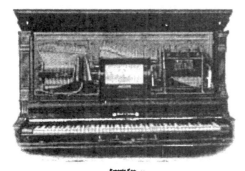

FIGURE 1.1 Poster from the early 1900s advertising a player piano using pneumatic control.

- maintainability
- cost

ALC is an important part of manufacturing systems especially in the areas of automation and robotics. In many cases the control system can be directly connected to the controlled pneumatic power system simplifying the system design. Air logic control is capable of performing many tasks and can be beneficial for control of many manufacturing processes. Superior control characteristics, ease of installation, high reliability, and low system cost can often be realized.

1.3 CONTROL CHARACTERISTICS

ALC is characterized by high energy output signals and slow response time. Where high levels of energy are required, the control signal can be amplified using the same air source multiplying the control signal to the required pressure. Response time of air controls is slow when compared to electronic signals but is much more compatible with systems when response time near that of the human operator is required.

Air logic control systems may be either analog or digital or a combination of both. Analog systems respond to any pressure signal between two specified extremes of a control range, and the control output may be modulated to convey the amplitude of the input signal as a control parameter. Digital control in air logic, as in other types of control systems, responds only to discrete changes in pressure and conveys only two control states: ON or OFF.

A great deal of air logic control is digital in nature resulting in simplified interaction with other digital control systems such as computers and PLCs. Some fluidic devices are analog in nature and provide variable signals which can be utilized for proportional control and modulation. Since the majority of industrial control systems use digital techniques, most of the discussion in this text will be devoted to digital control systems.

The primary disadvantage of ALC is that supply air for most control systems must be clean and dry. Some industrial applications preclude the use of ALC because of this requirement. Particulate matter and condensate can quickly clog passages in ALC devices and render a well-designed system inoperative and unreliable. This disadvantage can be easily overcome in most applications by proper fluid conditioning devices (Table 1.1).

TABLE 1.1
Characteristics of ALC

High Energy Output
Slow Response Time
Analog or Digital
MPL or Fluidic
Easy to Interface
Long Life
High Reliability
Good Accuracy
Excellent Repeatability
Good Resolution
Safe

TABLE 1.2
Types of ALC

MPL
Moving Parts Logic
uses moving parts (valves)

Fluidics
nonMPL Logic (no moving parts)
uses fluid flow phenomenon

1.4 MOVING PART LOGIC (MPL)

Most digital air logic control is accomplished with valving or devices with moving parts. This type of control is referred to as "moving part logic" or MPL. In MPL switching action is accomplished by opening and closing valves. Actuation of these valves can be manual, mechanical, electrical, or by fluid pressure (air or hydraulic). Miniaturization of valves and circuit manifolds for MPL applications allows high density circuits and fast operation.

1.5 FLUIDICS

Fluidics is a technology utilizing phenomenon of fluid flow in devices to accomplish control functions. These functions may be digital or analog. A great deal of fluidic technology is applied to sensing devices and provides a solution to problems which preclude the use of electrical type sensors such as hazardous area applications or any application where it is undesirable to have electrical currents present (Table 1.2).

1.6 FUTURE DEVELOPMENTS

Combining functions of MPL devices and further miniaturization will allow higher density controls in the future. Addition of electronic circuitry (chips) to MPL and fluidic devices will provide interesting hybrid controls with powerful capabilities. The availability of more sophisticated electronics communications is simplifying the interfacing of ALC with electronic and hydraulic systems making these combinations of control systems more attractive and less expensive.

The combination of fluidic devices, especially sensors, and solid state electronic switching devices will provide the designer with a wide variety of logic functions allowing the design of powerful control systems for the increasingly complex demands of automated manufacturing systems. Industrial computers, both the PC and PLC, have and will continue to provide a powerful interface between machines, humans, and ALC. The programmability of computers provides a very necessary flexibility for ALC and enhances the capability and applicability of both MPL and fluidic systems when used in ALC systems.

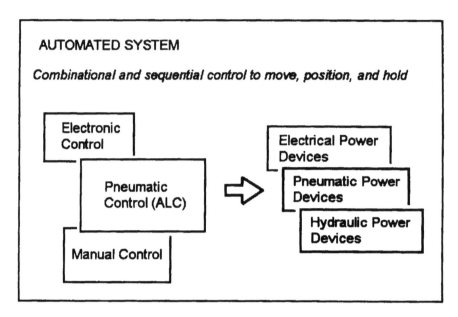

FIGURE 1.2   Automated system control

CHAPTER 1 REVIEW QUESTIONS

1. Define ALC.
2. What are some advantages of ALC?
3. What are some disadvantages of ALC?
4. What are some characteristics of ALC?
5. Define analog.
6. Define digital.
7. Define MPL.
8. Define fluidics.
9. Why would you choose MPL in place of fluidics?
10. How can electronic controls be combined with ALC?
11. What are some advantages of electronic controls over ALC?
12. Identify some applications of ALC in your plant.

2 Properties of Compressible Fluids

Our atmosphere provides a convenient fluid for energy conversion and transmission. The availability of air is unlimited and the cost of this fluid is inexpensive (politicians have not figured out how to tax us for it yet). Since air is a compressible fluid, predicting its behavior under varying conditions can become a complex task. Fortunately a great deal of information has been accumulated on this subject, so we need not reinvent it, only put it into convenient and useful form. Two types of problems may be encountered in the design or application of any pneumatic system: static and dynamic. Static properties of air are presented with the objective of providing a means for predicting pressure and forces developed in a system under any conditions of pressure and temperature. Dynamic properties are useful to predict pressure losses and available capacity of air flow.

2.1 STATIC PROPERTIES

Air is a solution of several gasses (78% nitrogen, 21% oxygen, 1% argon, carbon dioxide, and others by volume) having a density of .0752 pounds/cubic foot at 1 atmosphere pressure (14.7 psia), 68F, and 0% RH. Atmospheric air also contains water vapor (humidity). Density varies directly with pressure and relative humidity and inversely with temperature. As air is compressed, the relative volume of water vapor which is incompressible is increased resulting in saturation and separation of water. The pressure and temperature at which condensation occurs is referred to as dew point. Pressures and temperatures below the dew point result in liquid water separating from the compressed air. To ensure proper operation of any pneumatic or air logic system, this condensate must be removed.

2.2 GAS LAWS

Kinetic theory shows us that material properties are due to the energy levels of the molecules of the material. As the energy level of molecules increase, they become more active and disassociate themselves from each other producing state changes from solid to liquid to gas. This molecular activity is also responsible for temperature and pressure changes.

Pressure can be described by the impaction of molecules on a container's walls in which the molecules are contained. Forces of impact affect the walls of the container by the momentum principle. As more molecules are introduced into the container, then it follows that the pressure in the container will increase: more molecules, more impacts (Figure 2.1).

TABLE 2.1
Definitions of Symbols Used in Equations 2.1–2.31

Symbol	Definition
A	cross sectional area (inches²)
ACFM	flow at working pressure and temperature (actual cubic feet/minute)
C	a constant
CFM	volumetric flow (cubic feet/minute)
CR	compression ratio (pressure in pipe/atmospheric pressure)
K	ratio of specific heat at constant volume to specific heat at constant pressure
L	pipe length (feet)
MW	molecular weight
N	rotational speed (RPM)
P	absolute pressure (psia)
Q	volumetric flow (CFM)
R	universal gas constant (53.3 ft-lbs/lb-°R for air)
Re	Reynolds number
SCFM	flow at standard conditions (standard cubic feet/minute)
T	absolute temperature (° Rankine)
V	volume (feet³)
W	weight (lbs)
Ẇ	weight flow (lbs/second)
c	flow coefficient
d	inside diameter of pipe or tube (inches)
f	friction factor
g	acceleration due to gravity (32.2 feet/sec²)
m	mass
ṁ	mass flow
n	number of molecules or moles (as specified)
w	weight (pounds)
v	velocity (feet/sec)
v_s	velocity of sound
ΔP	pressure drop (psi)
ρ	mass density
μ	absolute viscosity (poise)
γ	weight density (lbs/foot³)

Based on this premise then:

$$P \propto n \times K \times T \tag{2.1}$$

The gas temperature is a result of the molecular activity or energy level. If proper units are used then:

$$P = \frac{n \times R \times T}{V} \tag{2.2}$$

TABLE 2.2
Properties of Dry Air

Property		Value	Units
Density at 30°F	0 psi	.0811	lbs/ft³
	80 psi	.522	
	100°F 0 psi	.0709	
	80 psi	.457	
Molecular weight at 68°F		28.952	
Gas constant		53.36	ft–lb/lb–°R
Thermal conductivity at 80°F		.01516	btu/hr–ft–°F
Cp/Cv mean ratio at 20°C		1.401	
Specific heat at constant pressure, 80°F		.2404	btu/lb–°F
Specific enthalpy at 80°F		153.2	btu/lb
Absolute viscosity at 80°F		.1214	lb/ft–s × E4

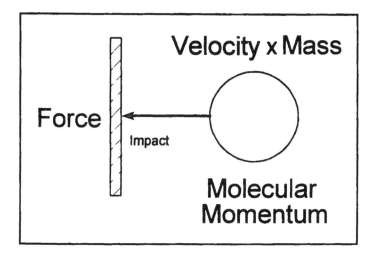

FIGURE 2.1   Molecular momentum

and:

$$n = \frac{m}{MW} \tag{2.3}$$

Equation (2.2) is more useful for pneumatic work if moles and mass are replaced with weight as follows:

$$P = \frac{w \times R \times T}{V} \tag{2.4}$$

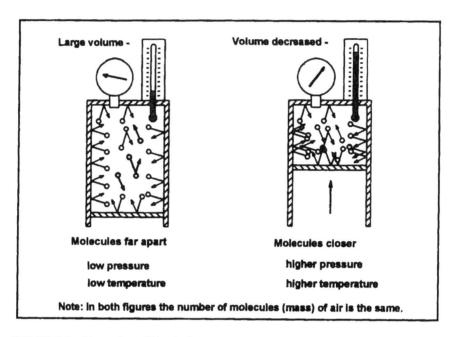

FIGURE 2.2   Illustration of kinetic theory

For any ideal gas:

$$w \times R = C \qquad (2.5)$$

This relationship becomes useful and is convenient to predict properties for air since R, the universal gas constant, is simply 53.3 ft-lbs/lb-°R for air. Substituting in (2.2):

$$C = \frac{P \times V}{T} \qquad (2.6)$$

For changing conditions of the same gas then:

$$\frac{P_1 \times V_1}{T_1} = \frac{P_2 \times V_2}{T_2} \qquad (2.7)$$

Equation (2.7) is the well-known combined gas law. When using the gas laws, remember pressure and temperature must be in absolute units (psia and °R). For any conditions pressure, volume, or temperature can be determined using (2.4), (2.6), or (2.7). Forces developed by calculated pressure can be determined by:

$$F = P \times A \qquad (2.8)$$

This simple relationship is useful to determine the force developed by any pressure actuated device (Figure 2.2).

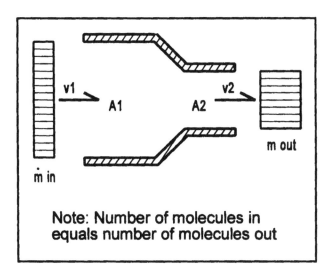

Note: Number of molecules in
equals number of molecules out

FIGURE 2.3   Illustration of continuity principle

2.3   DYNAMIC PROPERTIES

Understanding the behavior of pneumatic devices requires an understanding of the motion of compressible fluids, specifically air. In most cases this motion can be considered to be one dimensional and can be characterized as flow. Flow can be steady state or transient. The discussion in this text will be limited to steady state flow since most analyses are required at steady state conditions. Transient flow involves changing parameters, and in most cases, can be neglected. Transients can be troublesome when they exceed the allowable tolerances of any parameter. To avoid transient problems, the system should be designed with a minimum of transient sources.

Steady state flow can be characterized by the continuity principle. The continuity principle states that the mass of fluid entering a system must equal the mass of fluid leaving the system. This statement must be qualified as being true for steady state flow only if the fluid is compressible as is air. Mathematically the continuity principle can be stated as:

$$\dot{m}_{in} = \dot{m}_{out} \tag{2.9}$$

where: m_{in} is mass flow in the system
         m_{out} is mass flow out of the system

Mass flow is defined as the mass of fluid passing through a control point in a system with respect to time. Mass flow can be expressed as:

$$\dot{m} = \rho \times v \times A = \frac{P}{R \times T} \times A \times v \tag{2.10}$$

Combining (2.9) and (2.10), the continuity principle becomes:

$$\rho_1 \times v_1 \times A_1 = \rho_2 \times v_2 \times A_2 \tag{2.11}$$

where the variables are as defined previously. Subscript 1 refers to inlet or station 1 parameters, and subscript 2 refers to outlet or station 2 parameters (Figure 2.3).
Since weight is:

$$W = m \times g \tag{2.12}$$

Then flow of fluid based on weight can be determined by:

$$\dot{W} = \dot{m} \times g \tag{2.13}$$

Most flow problems involve volumetric flow so the relationship between mass or weight flow needs to be defined. Customarily:

$$Q = \frac{\dot{m}}{p} = \frac{\dot{W}}{\gamma} \tag{2.14}$$

Customary engineering units of air flow are cubic feet per minute (CFM). Since volumetric flow changes with pressure and temperature and it would be difficult to specify flow characteristics for a wide range of these parameters, a standard set of conditions needs to be defined for specifying flow. Flow at specified pressure and temperature is referred to as actual or ACFM. Flow rate at standard conditions is referred to as standard or free flow (SCFM). Standard conditions are defined as 1 atmosphere (14.7 psi), temperature of 68°F, and a relative humidity of 36% (Figure 2.4).
The conversions for ACFM and SCFM are:

$$SCFM = ACFM \times \frac{P + 14.7}{14.7} \times \frac{528}{T + 460} \tag{2.15}$$

and:

$$ACFM = SCFM \times \frac{14.7}{P + 14.7} \times \frac{T + 460}{528} \tag{2.16}$$

Flow parameters through piping, fittings, and valves have been determined by testing, and simplified relationships have been developed based on test data. Manufacturers publish test data and aids for calculating flow characteristics for the devices they manufacture. A standardized flow parameter for valves and other devices is the flow coefficient or C_v. Cv is determined empirically by measuring flow, pressure drop, and temperature through a device and calculating by:

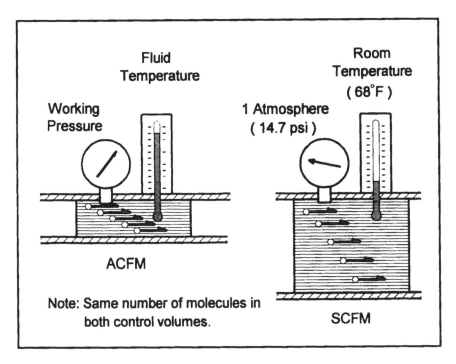

FIGURE 2.4   ACFM and SCFM comparison

$$Cv = \frac{SCFM}{22.48} \cdot \sqrt{\frac{T}{(P1 - P2) \cdot P2}}$$

(2.17)

where: P1 is inlet (upstream) pressure
              P2 is outlet (downstream) pressure

To determine flow through a valve use:

$$Q = 22.48 \cdot C_v \cdot \sqrt{\frac{(P_1 - P_2) \cdot P_2}{T}}$$

(2.18)

If:

$$P_2 \leq .53 \, P_1$$

(2.19)

P_2 is said to be critical and (2.18) can be simplified for all $P_{2critical}$. Then:

$$Q_{max} = 13.61 \times P_1 \times C_v \times \sqrt{\frac{1}{T}}$$

(2.20)

(2.20) is based on the relationship for the limiting velocity of a gas in a system being:

$$v_s = \sqrt{K \times g \times R \times T} \qquad (2.21)$$

For air (2.20) is simplified by substituting:

K = 1.4
g = 32.2 ft/sec²
R = 53.3

The limiting velocity for air which is the speed of sound can be calculated by:

$$v_s = 49 \times \sqrt{T} \qquad (2.22)$$

which is 1087.4 ft/sec at 32 F.

Pressure drop in a system can be determined by several techniques. The simplest relationship pressure drop in air lines is the Harris formula:

$$\Delta P = \frac{c\, L\, Q^2}{CR\, d^5} \qquad (2.23)$$

For schedule 40 commercial grade pipe c can be determined by:

$$c = \frac{.1025}{d^{.31}} \qquad (2.24)$$

Substituting this value of c into 2.23 results in:

$$\Delta P = \frac{.1025\, L\, Q^2}{CR\, d^{5.31}} \qquad (2.25)$$

Work required to determine pressure drops can be minimized by using manufacturers' data which is included in product brochures. Most manufacturers include tabulations or graphs of ΔP data for ease of application of their product. If values of c or manufacturers' data are not available, more basic, but much more complex, calculations must be used. The following equation can be used to determine pressure drop for any flowing gas assuming constant temperature at inlet and outlet:

$$P_1^2 - P_2^2 = \rho_1\, V_1^2 P_1\left[\frac{fL}{D} - 2\, \ln\frac{P_2}{P_1}\right] \qquad (2.26)$$

where f must be determined based on Reynolds number from:

$$R_e = \frac{\rho V D}{\mu} \tag{2.27}$$

Using this value of Reynolds number then f can be determined by:

$$f = \frac{64}{R_e} \tag{2.28}$$

If Re $\leq$ 2000 (laminar flow) and

$$f = \frac{.316}{R_e^{\frac{1}{4}}} \tag{2.29}$$

If Re > 4000 (turbulent flow).

Note that Reynolds number between 2000 and 4000 is referred to as critical which translated means "we don't know." In this flow regime the flow is usually assumed to be turbulent to design for worse case condition.

Air consumption of a device during operation can be determined by:

$$Q_{cl} = \frac{disp \times speed}{1728} \tag{2.30}$$

For a rotating device such as an air motor, (2.24) becomes:

$$Q_{cr} = \frac{disp \times N}{1728} \tag{2.31}$$

2.4 SUMMARY

The discussion in this chapter has presented only a few important relationships for compressible fluid flow and pneumatic principles as needed for an understanding of the operation of air logic devices and systems. Further studies in compressible fluid mechanics should be pursued for more in depth understanding of compressed air properties. An understanding of the relationships presented will allow determination of numerical solutions to problems, and more importantly, will provide an appreciation of properties of compressed air and its application.

CHAPTER 2 REVIEW QUESTIONS AND PROBLEMS

1. Explain the difference between static and dynamic properties of air.
2. What are the major elements present in air?
3. Why is relative humidity a concern to compressed air systems?
4. Explain kinetic theory.

5. Define pressure using kinetic theory.
6. An air receiver contains 5 pounds of air. The dimensions of the receiver (inside) are 2 feet in diameter and 3 feet long. What is the pressure of the air in the receiver assuming the temperature to be 80°F?
7. If the temperature of the air in problem 6 is increased to 100°F, determine the pressure at this temperature.
8. A system has a pressure of 200 psi and its temperature changes from 60°F to 110°F. Determine the new system pressure.
9. A 2" diameter cylinder is extended 24" with an air pressure of 80 psi. If the rod is retracted 6" by applying a load (assuming no air can escape from the system and the temperature change is negligible), what will the new pressure be?
10. What force will be developed by a 2" diameter cylinder extending with 80 psi pressure?
11. What pressure will be developed in a 2" diameter cylinder if a 100 pound force is applied to the cylinder rod (assuming no air escapes and the temperature change is negligible)?
12. State the continuity principle.
13. Determine the maximum flow that can be developed in a .200" I.D. tube.
14. The isolation valve in an air system has a Cv of 1.4. During system operation, a manometer shows a .5 psi drop across the valve. Assuming the temperature to be 80°F, determine the flow through the valve.
15. Convert 10 ACFM at 100 psi, 80°F to SCFM.
16. Convert 10 SCFM to ACFM at 100 psi, 80°F.
17. What is the maximum velocity air can attain in a system at 100°F?
18. What pressure drop will be developed by flowing air at 10 SCFM through 100 feet of 1" schedule 40 pipe at 80°F?
19. An air motor having a volumetric displacement of .8 cubic inches is running at 2000 RPM. Determine the air consumption for this motor.
20. What is the critical pressure for shop air flowing through a device?
21. What is Reynolds number? Why is it necessary to determine Reynolds number?
22. Explain the meaning of Reynolds number between 2000 and 4000.
23. Determine the flow coefficient for a valve which develops a .5 psi drop at 80 psi inlet and 80°F air temp with flow of 10 SCFM.
24. Determine the pressure drop developed by the valve in problem 23 if the flow is increased to 12 SCFM.

3 Fundamentals of Pneumatics

Most manufacturing operations use compressed air for a variety of functions. Air is a convenient fluid for energy conversion and transmission as well as control. The availability of air is unlimited and the cost of this fluid is inexpensive. Once used the air can be exhausted back into the atmosphere. Using air in pneumatic systems is non-polluting and does not harm the environment. Since air has a very low density, low inertia forces are developed in moving the fluid allowing high accelerations and decelerations. Also the viscosity of air is low, minimizing system pressure loss. Because air is compressible, precise positioning and velocity control are not possible. Pneumatic system pressures compared to hydraulic systems are quite limited. Hydraulic systems commonly use 10,000 psi as compared to high pneumatic pressure of 250 psi.

3.1 COMPRESSORS

Compressors are energy conversion devices which produce pressure energy by reducing the volume of intake air at low pressure and expelling the air at a higher pressure. The function of the compressor is to deliver a sufficient quantity of air at a desired pressure. Two types of compressors commonly used are the diplacement type and the dynamic type.

Displacement type compressors produce pressure increase in air by confining a volume of air to a decreasing volume. The most common displacement compressor is the reciprocating piston compressor. The dynamic compressor produces pressure increases in air by introducing kinetic energy into a volume of air through the use of an impeller or a propeller. The compressor mechanism simply moves the air molecules at an increased velocity producing an increase in pressure energy. Most industrial compressors are displacement type and most commonly, the reciprocating piston type. The output of a compressor is usually expressed in units of rate of volume displaced (CFM). Output is determined by the piston size, stroke, speed, and number of pistons if a positive displacement compressor (Figure 3.1).

Dynamic type compressors are suited for applications requiring large volumes of air, but large variations in pressure result with changing output demands. During compression, heat is developed (refer to the Combined Gas Law in Chapter 2) which can result in equipment damage if not removed. Most compressors are air cooled by convective fins surrounding the piston cylinders.

3.2 RECEIVERS

It is convenient to provide compressed air from a centralized source which is supplied by a compressor having sufficient capacity to act as an infinite supply. Since demand on the system will never be consistent, a storage device must be provided to allow

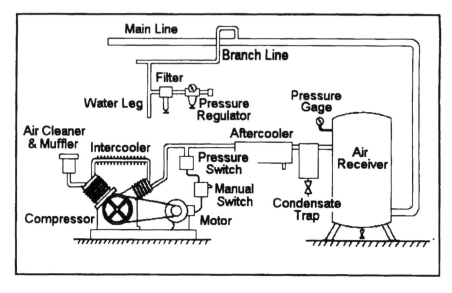

FIGURE 3.1   Typical shop air system using a displacement compressor

TABLE 3.1
Definitions of Symbols Used in Equations 3.1–3.6

Symbol	Definition
A1	area of low pressure piston (in^2)
A2	area of high pressure piston (in^2)
Ap	piston area (in^2)
Ar	cross sectional area of rod (in^2)
D	outside diameter (inches)
F	force generated (pounds)
F1	force generated by low pressure piston (pounds)
F2	force generated by high pressure piston (pounds)
FS	safety factor
Pmax	maximum pressure delivered by the compressor (psi)
Pmin	minimum pressure delivered by the compressor (psi)
P	pressure (psi)
Pb	burst pressure (psi)
Pw	working pressure (psi)
Qr	consumption rate of system (cfm)
Qc	output of compressor (cfm)
S	ultimate tensile stress (psi)
t	wall thickness (inches)
T	time receiver can supply required air (mins)
Vr	receiver volume (cubic feet)

the compressor to maintain sufficient supply of energy when required and remain idle when demand is minimum. The storage device is called a receiver and is simply a tank of sufficient capacity to hold compressed air from the compressor as it operates, and to release the air to the system as it is needed.

The size of the receiver may be determined by the following:

$$Vr = \frac{14.7\ t\ (Qr - Qc)}{Pmax - Pmin} \tag{3.1}$$

The receiver also serves to dampen out pressure transients from the compressor and from the devices using the pneumatic system to provide air at a constant pressure.

The proper operation of an air compressor system requires:

1. Sufficient power to the compressor
2. Sufficient air flow from the compressor
3. Compensation for system leakage
4. Sufficient pressure to operate devices in the system
5. Compensation for pressure drops in the system
6. Correct temperature of air and coolants

3.3   PRESSURE SWITCH

The operation of a compressor is controlled by a pressure switch which senses the pressure of the air in the receiver and, at a minimum pressure in the tank, will switch the compressor on. At some maximum pressure in the tank, the pressure switch turns the power off. Setting the maximum and minimum pressures can be tricky since the adjustments on the switch:

- raise or lower the range
- vary the differential between switch on and switch off

Most operations using a central compressor system produce shop air up to 120 psi and with minimum system pressure of as low as 60 psi. Normally the design value for shop air is 80 psi (average). When developing a system, one must be cautious of system demand causing pressures falling below 80 psi. System pressures may inadvertently fall as low as 60 psi if long lines are used and heavy demand is placed on the system. It is usually a good idea to place a recording pressure monitor in the line at the point of use and record pressure changes at periods of high demand if pressure is critical to proper operation so the system may be designed for the minimum pressure constraint.

3.4   PNEUMATIC SYSTEMS

Compressed air usually requires some amount of conditioning to provide a satisfactory fluid medium. Conditioning involves regulation, filtering, lubrication, drying,

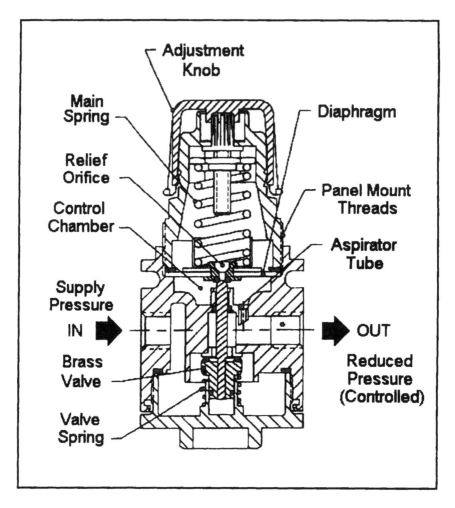

FIGURE 3.2   Air line pressure regulator (Courtesy of Numatics, Inc., Highland, MI).

and muffling. Regulators provide a constant pressure to the system by modulating the air flow to the system. Most regulators use a spring loaded diaphragm which limits the air flow as the pressure in the line approaches the set value. Keep in mind the regulator can regulate down from the inlet pressure but not up. Also most regulators require flow in the system to regulate to the set pressure (Figure 3.2).

Air is compressed by mechanical means and flows through mechanical devices which may contain many contaminants. These contaminants can be of sufficient size to block orifices in valves and other equipment especially miniature air logic devices. Filters are placed in line to remove particles in size down to 25 microns. By nature of the flow characteristics of the filter, some amount of moisture is removed but not all and, in some applications, not enough. Air is a very dry fluid with regard to lubrication, so equipment with moving parts operated by air requires some type of added lubrication for long life. Lubricators are commonly used in systems which inject a fine mist of oil into the air stream to provide needed lubrication.

3.5 NOISE

Noise in pneumatic systems is caused primarily by the compressor and by escaping (exhaust) air. Compressor noise can by reduced by:

- Adding a muffler on the air inlet
- Enclosing the compressor in a sound deadening box
- Locating the compressor in a remote location

Noise due to escaping air can be reduced by installing a muffler to the air exhaust. Mufflers must:

- Have a low resistance to flow
- Be corrosion resistant
- Attenuate the sound sufficiently

OSHA regulations specify the level of sound acceptable in the workplace. High pitched and loud sounds should be reduced in intensity to comply with acceptable OSHA levels. Excessive sound levels can produce stress in human behavior resulting in lowered productivity and liabilities.

3.6 FLUID CONDITIONING

Air logic devices may contain small orifices which are required for their intended function. Contamination in air passing through these devices can plug the orifices and interfere with their operation. Contamination may be in the form of particulate matter or entrained liquids.

Sources of contaminants can be the system's environment or the system itself. To ensure reliability of the system's operation, these contaminants must be prevented from entering the system or be removed from the air before exposure to critical components. Good system design will minimize contaminants in the system and provide equipment to remove any material that could produce problems. Devices are available which will, when properly applied, remove contamination and allow reliable system operation.

Typical contamination sources include:

- Rust and scale from pipe
- Sealing materials used to join pipe
- Humidity in air
- Air borne particles from processes
- Scraping and grinding of parts in the system as they operate
- Lubricants added to the system
- Teflon tape

Strainers are coarse filters usually made from a screen having over 100 mesh (.0059 inch openings). Particles over 149 microns in size are separated by this size opening. The advantage of a strainer is that it produces little restriction to flow.

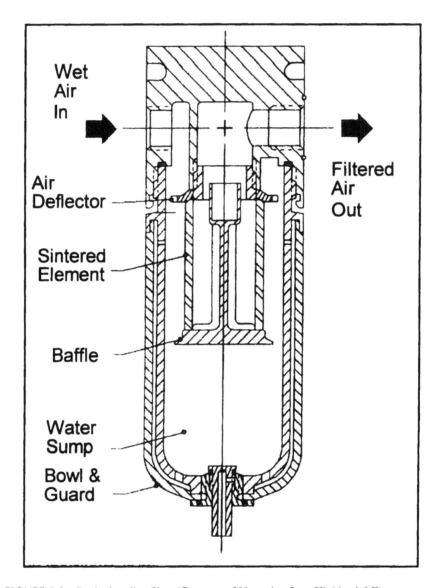

FIGURE 3.3   Particulate line filter (Courtesy of Numatics, Inc., Highland, MI).

Because of their inability to separate small contaminates, strainers are not recommended for air cleaning.

   A filter is used to extract contaminants from the air stream by flowing the air through a porous medium and by mechanical action (Figure 3.3). Filters should be selected based on:

*   Pressure rating
*   Port size
*   Particle size removed

- Pressure drop
- Ability to remove oil and water
- Service required

Coalescing filters carry air flow from the inside out (opposite the flow pattern of the standard particulate filter). Contamination is captured in the filter matrix and collected into larger droplets as it collides with the glass microfibers. A drain layer collects the larger droplets and gravity flows the contamination to the sump in the bottom of the filter housing. Coalescing filters can remove as much as 99.97% of contaminants as fine as .3 microns. For optimum performance it's a good idea to have a pre-filter in line before the coalescing filter to remove larger materials and extend the operating life of the filter (Figure 3.4).

When very dry air is required, desiccant dryers may be used. Desiccant dryers are similar to particulate dryers except the filter medium is usually silica gel or alumina which adsorbs moisture in the air as it flows through the media. A pre-filter which will remove the majority of the moisture should be placed before the dryer to ensure long service life of the dryer.

3.7   FLUID TRANSMISSION

Fluid signals are transmitted through conduits. Three types of conduits are pipe, tubing, and hose. The selection of conduit type is based upon several factors including pressure of signals, routing, environmental conditions, flow requirements, devices in the system, and permanence of the system. Part of the fluid transmission process includes conditioning. Because of small orifices and passages in air logic controls, it is vital that air used in their operation be clean and dry.

Pipe

Pipe refers to a rigid conduit usually with a relatively thick wall. Pipe may be welded, seamless, or extruded types.

Welded pipe is manufactured by forming (rolling) a flat strip of metal into a cylinder and welding the seam to produce a cylinder of nearly the size required, then finishing the cylinder by additional rolling into the required dimensions for the pipe. Seamless pipe is manufactured by hot forming a metal ingot through consecutive rollers reducing the cross section several times until the desired cross section is attained. At final rolling, a mandrel is used to pierce the metal to form the hole in the pipe. Extrusion usually refers to processing of plastic pipe. By extrusion, plastic is heated and, under heat and pressure, it is flowed through tooling to produce the desired cross section for the pipe. Pipe is specified by nominal size and schedule.

Tubing

Tubing can be rigid or flexible depending upon wall thickness, material, and the process by which it is manufactured. As pipe types, tubing may be welded, seamless, or extruded and may be made of either metals or plastic. The primary difference between tubing and pipe is that tubing is manufactured to a specific outside diameter.

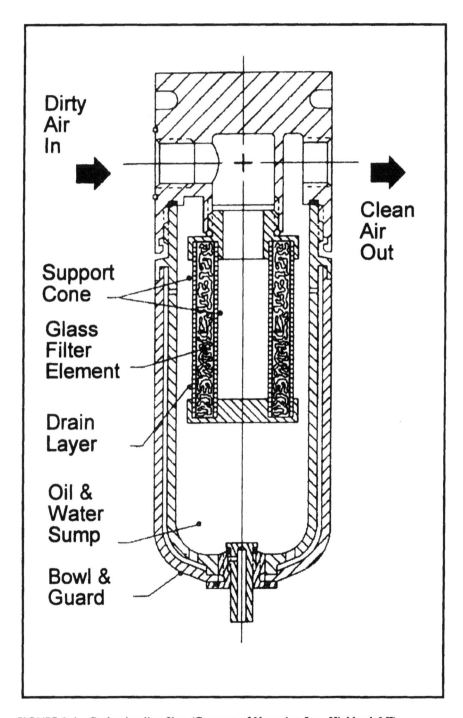

FIGURE 3.4 Coalescing line filter (Courtesy of Numatics, Inc., Highland, MI).

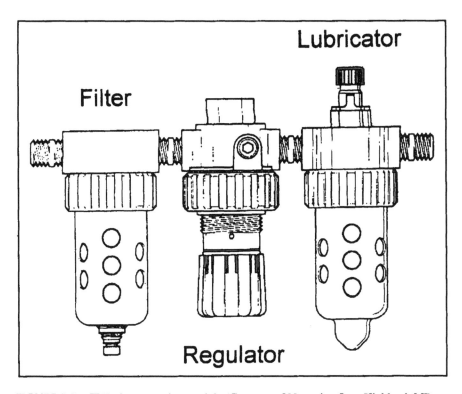

FIGURE 3.5   FLR air preparation module (Courtesy of Numatics, Inc., Highland, MI).

Some welded tubing is available in long lengths in coils (up to 1000 feet) and can be bent to route it rather than use fittings. Common metals used for tubing include low carbon steel, stainless steel, and copper. Common plastics used for tubing for pneumatic service are nylon, polyethylene, and polyurethane.

Hose

Hose is used where flexibility and high working pressures are required. Hose is manufactured by wrapping layers of reinforcing materials over an extruded thin wall plastic tube and then extruding a protective elastomeric jacket over the composite. The reinforcing material can be metal wires, fiber yarns, and plastic tapes.

3.8   FITTINGS

Fittings serve to fasten lines together and to fasten devices to lines. Type of fitting used depends upon the type of line being connected, the permanence of the connection, and the required integrity of the connection. There are many fitting styles and designs available as stock products. Some designs are specified for specific applications so caution should be used when specifying components. Details of the application are important considerations when engineering the system and must match the intended application details of the component.

3.9   MANIFOLDS

The manifold functions to provide a common connection point for multiple lines. Manifolds can save a great deal of space and can simplify routing of lines. Standard type manifolds are available and most manufacturers will custom make manifolds to suit your needs. A manifold can simply be a machined steel block with a central cavity and ports tapped off from the cavity. More complex manifolding may be cast to provide complicated cavities and porting arrangements.

Manifold subplates are available which have integral circuit connections for basic functional circuits. These subplates simplify construction of circuits and provide neat arrangements of circuit elements. They are especially useful for multiple circuits.

3.10   PRESSURE RATING

The critical characteristic for pressure pipe, tube, or hose is its burst pressure. Burst pressure can be calculated by Barlow's formula:

$$P_b = \frac{2St}{D} \tag{3.2}$$

It is never desirable to apply pressures near the burst pressure to a line since this can result in failure or at least permanent deformation of the line. To determine safe working pressure, Equation 3.2 is modified as:

$$P_w = \frac{2tS}{FS\,D} \tag{3.3}$$

A factor of safety of 4 is acceptable for normal operation. If severe pressure shocks or pulses occur, a factor of safety as high as 10 may be required. All operating conditions need to be considered when selecting an appropriate factor of safety.

3.11   SIZING

Sizing pipe, tube, and hose is usually an exercise in compromise. For least pressure drop and greatest flow capability, the larger the diameter, the better; however, as tubing diameter goes up, burst pressure capability goes down. Also as diameter goes up, so does cost. So compromise is necessary to select the optimum tube size.

Remember:

Pipe: pipe size is specified by nominal size and schedule
Tube: tube size is specified by outside diameter (OD) and wall thickness
Hose: hose size is specified by inside diameter (ID) and maximum operating
   pressure.

3.12 MATERIALS SELECTION

When selecting materials for pipe, tube, or hose, the following must be considered (See Table 3.2):

- Fittings that are appropriate: many constraints affect the appropriateness of fittings to be used. Some important considerations include vibration the system can be exposed to, technical expertise of personnel that may be working on the system, cost constraints, and availability.
- Operating pressure: high pressure lines, of course, require metal tubing. Before a material is selected, refer to the manufacturer's burst pressure data or calculate burst pressure using Barlow's formula. Include sufficient safety factor when determining a safe working pressure. Be sure to consider pressure pulsations and abnormal pressure variations.
- Operating temperature: elevated temperature decreases the yield stress of materials especially plastics. If the lines in a system may be exposed to temperature variations, burst pressure should be determined based on the yield stress of the material at the maximum elevated temperature the system may be exposed to. Extreme low temperatures also have dramatic effects on materials especially plastics. If the system is exposed to subzero temperatures, plastics must be chosen carefully to avoid use of materials which become brittle at the working temperatures.
- Flexibility: flexibility of the tabing may be a concern for three reasons (1) if the lines must be routed by bending around areas during installation; (2) if lines are moved or removed for system maintenance or upgrading; and (3) if the lines are moved as a part of the system operation. Depending upon the severity of these concerns, hose or flexible tubing may be necessary.
- Cost: system cost may be affected adversely if line quality is sacrificed for cost. The best product for the application must be selected to optimize ultimate costs.
- Chemical compatibility: industrial environments may subject a conduit material to adverse substances. The compatibility of materials is extremely specific, so each material must be considered based on historical experiences and compatibility tests.
- Flow requirements: material selection may affect the flow through the tube due to friction factor of the wall. Friction factor varies with material and also with processing. The smoother the wall, the lower the frictional loss will be during flow.
- Length requirements: available lengths are determined by the manufacturing process used to manufacture the conduit. Usually pipe is limited to random lengths which are between 12 and 22 feet long.

3.13 PNEUMATIC ACTUATORS

The usual function of a pneumatic power system is to produce controlled motion. To accomplish this function, attributes which must be controlled include:

TABLE 3.2
Material Selection Considerations for
Pipe, Tubing, and Hose

Fittings that are Appropriate
Operating Pressure
Operating Temperature
Flexibility
Cost
Chemical Compatibility
Flow Requirements
Availability
Length Requirements

- Force developed
- Velocity of motion
- Acceleration of motion
- Position

Typical actuators designed to accomplish these tasks are:

- Cylinders
- Rotary actuators
- Air motors
- Slides
- Grippers

Cylinders

Cylinders provide linear motion. The cylinder converts pneumatic pressure to mechanical force. The cylinder consists of a cylindrical body housing a sliding circular piston with a rod attached. Pressure on the piston surface develops proportional to the piston area. Force extending can be determined by:

$$F = P \times A_p \qquad (3.4)$$

When retracting, the area of the rod does not produce any thrust, and so must be subtracted from the piston area:

$$F = P \times (A_p - A_r) \qquad (3.5)$$

Similar mathematical models can be used to determine velocity and displacement. A variety of models of cylinders are available with different styles of mounting hardware and rod connections. Standard sizes are based on piston diameters and stroke lengths (Figure 3.6).

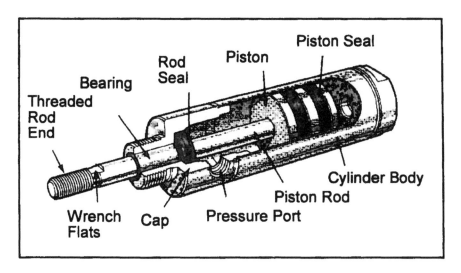

FIGURE 3.6   Cut away view of a typical air cylinder (Courtesy of ICI Norgren, Inc., Rockford, IL).

Rotary Actuators

The rotary actuator converts pneumatic pressure to rotary mechanical torque. Rotary actuators may utilize integral cylinders which produce linear motion of a toothed rack which rotates a pinion producing rotary motion. Actuators are also available which produce torque by means of a vaned rotor. Torque developed is a function of the actuator vane area and pressure applied. Rotary actuators produce only partial rotation in both directions but never a complete revolution (Figure 3.7).

Air Motors

Air motors produce rotation by the application of pressure on a multivane rotor. Torque produced is a function of the blade area and the pressure applied. Air motors are available in sizes based on their power output. Power outputs from 1/10 hp up to 10 hp when operating at 100 psi inlet pressure. Free (no load) speeds as high as 15,000 ROM are available. The speed of an air motor is dependent upon applied load. As torque load is applied, the speed slows down and as load is decreased, the motor speed increases. Air consumption of air motors is high, and these devices are relatively inefficient.

Slides

Slides are specialty cylinders which produce linear motion that is torque resistant and precisely normal to the direction of generated motion. Slides are useful for producing long linear strokes which are free of any rotation.

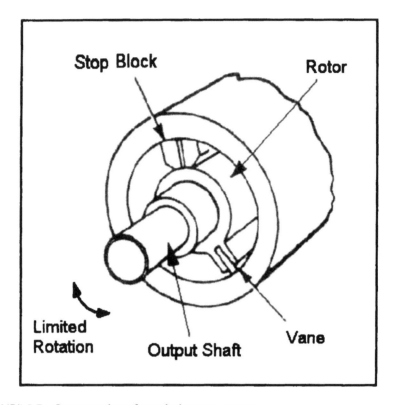

FIGURE 3.7   Cut away view of a typical rotary actuator

Escapements

Escapements are pairs of cylinders in a common housing that will isolate an individual part, then release it. Figure 3.8 illustrates this operation. In Step 1, the part is separated from the others in que by extending (EXT) fingers x and y. During Step 2, fiber y is retracted (RET) allowing the part to escape while holding back those in que. In Step 3, both fingers are extended still holding back the parts in que until Step 4, when finger x is retracted to allow another part to be isolated when Step 1 is repeated. Escapements are especially useful in gravity slide applications.

Grippers

Grippers are specialized cylinders designed to apply forces to a product for the purpose of holding a product while moving it. Two finger grippers are most common. Two styles are available: angular and parallel. Angular grippers are least expensive and develop the fastest gripping motion with least weight. Parallel grippers develop exact parallel gripping motion. Parallel motion comes at a higher cost when compared to angular motion developed by angular grippers (Figure 3.9).

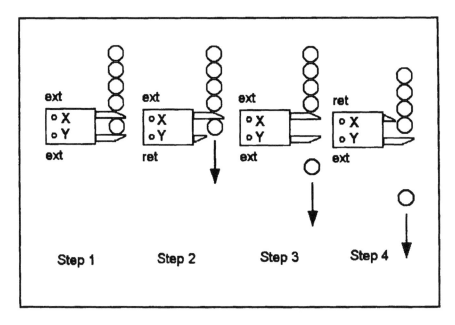

FIGURE 3.8   Operating steps of an escapement

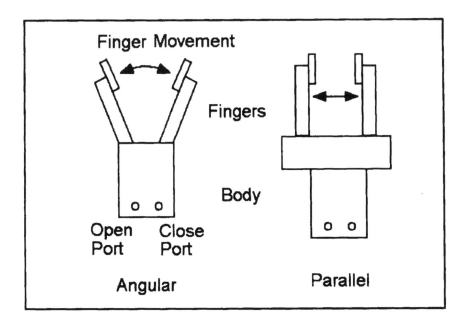

FIGURE 3.9   Operation of angular and parallel grippers

3.14   INTENSIFIERS

An intensifier is used to increase fluid pressure. The intensifier has two pistons, a high pressure piston and a low pressure piston which are mechanically coupled. Since the force developed on the low pressure piston (F1) is directly transferred to the high pressure piston (F2) through a common shaft, the force F2 must equal F1.

This relationship can be modeled by:

$$F1 = F2$$
$$P1 \cdot A1 = P2 \cdot A2$$
$$P2 = P1 \cdot \frac{A1}{A2} \qquad (3.6)$$

Intensifiers can be purchased or can be constructed by connecting the rods of two cylinders having the appropriate diameters.

3.15   CONTROLLING PNEUMATIC POWER

To produce the desired system function, control of the following four system parameters may be necessary:

- Direction of flow
- System pressure
- Air flow (velocity)
- Device sequence (timing)

The directional control valve (DCV) controls the fluid path in a circuit which determines the direction that a device travels. The DCV is used to extend or retract the rod of a cylinder or operate a motor CW or CCW. A variety of port arrangements are available allowing DCVs to be applied conveniently to most applications. The check valve allows flow in one direction but restricts the flow in the opposite direction. Because the check valve controls flow direction, it can be considered to be a simple DCV.

Valves can be operated by various types of actuators depending upon the type of control input available. Actuators include:

- Electrical solenoids
- Pneumatic cylinders
- Hydraulic cylinders
- Manual operators

System pressure can be controlled by regulators described previously. Miniature regulators are available which can be placed in systems to produce pressures at specific locations. Air flow is usually controlled by a modulating valve such as the needle valve which produces a nearly linear change in flow as the valve control is adjusted. Globe valves and ball valves do not provide this linear flow change. The flow control contains a needle valve with a check valve in parallel with it so flow only in one direction is controlled. Flow in the opposite direction free flows through the check valve. Sequencing of system functions can be accomplished by a variety of valves which may be simple two port solenoid valves or complex modulating servo valves depending on the required function.

3.16   SUMMARY

This chapter has presented an overview of pneumatic power fundamentals especially where the technology relates to air logic systems. It is not within the scope of this book to present details of pneumatic power circuit design. The designer, when faced with a project involving air logic, seldom has the luxury of dealing only with the control portion of the system and so must be knowledgeable in related technologies as well.

CHAPTER 3 REVIEW QUESTIONS

1. Define ALC.
2. What are some advantages of ALC?
3. What are some disadvantages of ALC?
4. What are some characteristics of ALC?
5. Define analog.
6. Define digital.
7. Define MPL.
8. Define fluidics.
9. Why would you choose MPL in place of fluidics?
10. How can electronic controls be combined with ALC?
11. What are some advantages of electronic controls over ALC?
12. Identify some applications of ALC in your plant.

4 Fundamentals of Logic Systems

Logic is the science of predicting the outcome of a series of inputs to a process. This outcome is referred to as the output of the process. By combining inputs in various manners, the output can be controlled sequentially or chronologically. Logic reasoning can be traced back to antiquity. Aristotle made the profound statement that "any statement can be either true or false but it can never be both and it can never be neither." This astute statement remained a philosophical curiosity until an English mathematician named Boole expounded upon it in mathematical terms, and from his work, the concepts of digital logic evolved and is now known as "Boolean algebra." The basis for Boolean algebra is a binary numbering system of only two values: 0 and 1. This abstract numbering system translates to real world conditions such as "false or true," "off or on," "no or yes," "not present or present," and "closed or open." The outcome of complex combinations of these conditions can be easily predicted by binary representation (digital logic) and algebraic manipulation (Boolean algebra).

4.1 SYMBOLISM

Digital logic is an abstract representation of conditions using combinations of 0s and 1s. These combinations must obey the basic laws of mathematics and may be operated on using algebraic manipulation to predict their outcome. Keep in mind that only two values or states exist (0 and 1) so the only outcome of combining 0 and 1 is 0 or 1. At first it may seem nonsensical to say the outcome of adding a 1 and a 1 is 1, but realize the only outcomes possible are 0 or 1. A more logical explanation of this process would be to consider 0 to be OFF and 1 to be ON. Then ON plus ON equals ON makes more sense. ON plus ON could not be OFF (Figure 4.1).

Therefore the symbols 0 and 1 should be read as "off or on," or "nonpassing or passing" rather than as their numerical values which may imply nonsensical results. MPL and relay symbolism will be used to present logic concepts in this chapter to provide a dualism which will provide real world illustrations to enhance the digital abstractions presented. The primary MPL device used will be a two position, two port block valve and single relay contact. The normal condition refers to the state of the valve with no signal pressure applied. Note that MPL or any valve is considered to be NC when no flow condition exists and NO when flow through the valve takes place. This is exactly opposite for relay logic (any electrical) device. NC switch or relay contacts allow current flow across the contacts and NO designates the no flow condition. To eliminate any confusion, both MPL and relay logic devices in our discussions will be referred to as 0 state (0) or nonpassing when no flow condition exists, and 1 state (1) or passing when flow is allowed. Logic relationships are symbolically represented by algebraic connectives. This algebraic representation uses the + to indicate "OR" (sum), the • to indicate "AND" (product), and ' (or ‾)

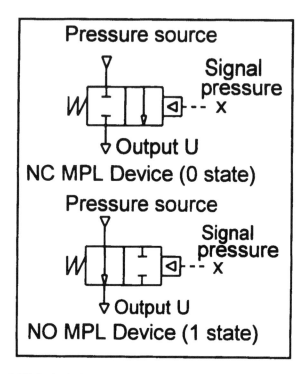

FIGURE 4.1   MPL logic device attached symbols

to indicate negation. Negation simply implies changing state. If a state is 0, then negating or changing that state can only result in the other available state or 1. Negating the 1 state results in the 0 state. Some physical examples of negation include an NC valve (0 state) held open or signal applied in the normal state to result in 1 state of the valve or a NO (1 state) valve held closed or signal applied in the normal state to result in 0 state of the valve (Figure 4.2).

4.2   BOOLEAN POSTULATES

If X is an input to a system and 0 and 1 are states as previously defined, then:

$$X = 1 \ \ if \ \ X \neq 0$$
$$X = 0 \ \ if \ \ X \neq 1 \tag{4.1}$$

This postulate states that if X is 1, then it is not 0 and if it is 0, it is not 1 (Table 4.1).

4.3   BASIC OPERATIONS

Basic operations in the first column of Equation 4.3 below are expressed as 1 AND 1 equal 1, 1 AND 0 equal 0, 0 AND 0 equal 0. The second column of 4.3, 0 OR 0

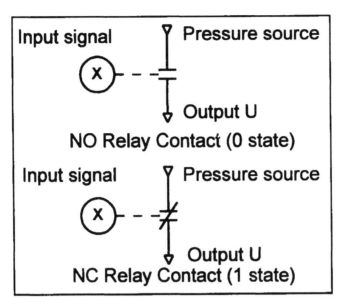

FIGURE 4.2   Relay logic device detached symbols

TABLE 4.1
Definitions of Symbols Used in Equations 4.1–4.29

Symbol	Definition
A, B, C	Designates a valve
NC	Normally closed or 0 state with no signal pressure applied
NO	Normally open or 1 state with no signal pressure applied
X, Y, Z	Signal pressure at port X, Y, or Z
U, V, W	Output at port U, V, or W
0	Nonpassing state or no signal pressure, OFF state
1	Passing or pressure at signal port, ON state
′	Negation
″	Double negation
+	Logic OR function
•	Logic AND function
=	"Results in"

---

equal 0, 0 OR 1 equal 1, 1 OR 1 equal 1. Note the AND operator corresponds to product (multiplication) and the OR operator corresponds to sum (addition).

$$
\begin{array}{ll}
1 \cdot 1 = 1 & 0 + 0 = 0 \\
1 \cdot 0 = 0 & 0 + 1 = 1 \\
0 \cdot 0 = 0 & 1 + 1 = 1
\end{array}
\qquad (4.3)
$$

Some authors use ∪ (cup) to indicate logic addition and the ∩ (cap) to indicate logic product. Negation indicates the inverse condition relative to the normal condition. If the logic statement is "the device is ON," the negation of this statement "the device is ON negated" means the device is OFF. Negation is stated by:

$$1' = 0 \;\; and \;\; 0' = 1 \tag{4.2}$$

which means the negated 1 is 0 and the negated 0 is 1 or the state is opposite the normal state.

4.4 COMBINING DEVICES

Several basic combinations of devices can be used to predict the output of any logic circuit. In nature, events occur in degrees of occurrence. Inputs may or may not produce an occurrence depending upon their amplitude. These types of systems are referred to as analog. Prediction and modeling of analog systems can be extremely complicated. To simplify systems to allow more easily predicted results and more easily understood system operation, digital systems are used. The concept of digital is simply a special case of analog in which an occurrence takes place when a discrete event takes place. In the digital system there is no high, low, almost, or degree of occurrence only on or off, true or false, yes or no. Industrial application of digital systems may require several inputs to control an output. Also many outputs may be necessary to provide process control. Combinations of inputs and outputs can result in complex systems. The degree of complexity may easily exceed any human's ability to comprehend the system's function or predict its operational characteristics. To allow analysis of a system's operation, the system can be subdivided into understandable elements and the operation of these elements combined for interpretation of the system's operation. Outcomes of the operation of elements can be illustrated by tabulating the possible element states along with the resulting outcome. This scheme of tabulating the logic element input and output states is referred to as a "truth table." Truth tables will be used throughout this book to illustrate logic element and simple system outcomes. A problem and a limitation to truth tables is they become extremely complex as the number of devices or elements in a logic system increases.

4.5 DEFINED FUNCTIONS

The postulates stated define the following basic functions:

YES

$$X = 1 \;\; then \;\; U = 1 \tag{4.4}$$

By definition if output is in 1 state when input is in 1 state, the function is a YES. Table 4.2 illustrates this logic function.

Example: A normally closed 2 port solenoid valve has no output with no input. When power (input) is applied to the solenoid there is output through the valve.

TABLE 4.2
Truth Table for YES Logic

INPUT	OUTPUT
X	U
0	0
1	1

TABLE 4.3
Truth Table for NOT Logic

INPUT	OUTPUT
X	U
0	1
1	0

TABLE 4.4
Truth Table for AND Logic

INPUT		OUTPUT
X	Y	U
0	0	0
1	0	0
0	1	0
1	1	1

NOT

$$X = 1 \quad then \quad U = 0 \tag{4.5}$$

For input of the NOT function there is no output (Table 4.3).

Example: A normally opened solenoid valve. When the solenoid is energized, there is no output through the valve.

AND

$$A \cdot B = U \tag{4.6}$$

The AND function is associated with multiplication (logic product). The function statement is read as the product of A AND B is the output U (Table 4.4).

TABLE 4.5
Truth Table for OR Logic

INPUT		OUTPUT
X	Y	U
0	0	0
1	0	1
0	1	1
1	1	1

TABLE 4.6
Truth Table for Negation and Double Negation

X	X′	X″
0	1	0
1	0	1

Example: Two normally closed solenoid valves are plumbed in series. Then power on solenoid of valve A is input A. Power on solenoid of valve B is input B. Output U is flow through valve A AND B.

OR

$$A + B = U \qquad (4.7)$$

The OR function is associated with addition (logic sum). Statement reads A OR B equals output U (Table 4.5).

Example: Two normally closed valves are connected in parallel. Power on the solenoid of valve A OR power on the solenoid of valve B produces an output U.

Negation

$$A' = U \qquad (4.8)$$

Negation involves use of the opposite states as the normal state (making the 1 state the normal state). Negated 0 is 1. Negation is signified by the ″ superscript.

Example: A normally opened solenoid valve has output with no input and when the solenoid is powered (input) there is no output.

Double Negation

$$X = X'' \qquad (4.9)$$

Negation implies changing state. If negated 0 is 1, then negating a second time will be 0 (original state) (Table 4.6).

TABLE 4.7
Truth Table for NOR Logic

INPUT		OUTPUT
X	Y	U
0	0	1
1	0	0
0	1	0
1	1	0

TABLE 4.8
Truth Table for NAND Logic

INPUT		OUTPUT
X	Y	U
0	0	1
1	0	1
0	1	1
1	1	0

NOR

$$A' + B' = U \qquad (4.10)$$

If the inputs to a system are changed to the negated state and they are plumbed in series, the output U will be the sum of the two inputs (Table 4.7).

Example: Two normally open solenoid valves are connected in series. The output is the flow through the valves when no inputs are present.

NAND

$$A' \bullet B' = U \qquad (4.11)$$

Two negated inputs connected in parallel will have an output U which is the product of the negated inputs (Table 4.8).

Example: Two normally open solenoid valves are connected in parallel. The output is flow through the valves until both are energized and changed to 0 state.

4.6 ALGEBRAIC EXPRESSIONS (BOOLEAN ALGEBRA)

It is important to realize that the logic we are dealing with uses two numbers (0 and 1). No others. Algebraic manipulation of relationships between these numbers is quite interesting and is useful to predict the behavior of logic devices and systems.

TABLE 4.9
Truth Table for Communitive Property

INPUT		LOGIC			
X	Y	X•Y	Y•X	X+Y	Y+X
0	0	0	0	0	0
1	0	0	0	1	1
0	1	0	0	1	1
1	1	1	1	1	1

4.7 MATHEMATICAL SYMBOLISM

The inputs to a digital logic device, as mentioned previously, have two states: off or on, closed or open, 0 or 1. The off (closed) state is represented mathematically by a 0. The on (open) state is represented by a 1. Inputs can be labeled with any symbol for conveyance. For the discussion on Boolean theorems or properties, the inputs will be designated by X, Y, and Z. The input may be in either state. Any other mathematical symbolism follows traditionally accepted convention.

4.8 PROPERTIES

Boolean theorems or properties are simply statements of the mathematical relationships that exist between the inputs when combined in two ways: AND and OR in different orders and combinations. Note that all of the properties have duals. A dual is obtained by changing all ANDs to ORs, changing all 0s to 1s, and visa versa.

Communitive

$$X + Y = Y + X$$
$$XY = YX$$
(4.12)

The order of the elements in a logic statement does not affect the result (Table 4.9).

Associative

$$(X + Y) + Z = X + (Y + Z)$$
$$(XY)Z = X(YZ)$$
(4.13)

The elements of a logic statement may be grouped in any order without affecting the result (Table 4.10).

Distributive

$$X + (YZ) = (X + Y)(X + Z)$$
$$X(Y + Z) = (XY) + (XZ)$$
(4.14)

TABLE 4.10
Truth Table for Associative Property

INPUT			LOGIC			
X	Y	Z	(X+Y)+Z	X+(Y+Z)	(X•Y)•Z	X•(Y•Z)
0	0	0	0	0	0	0
0	0	1	1	1	0	0
0	1	1	1	1	0	0
0	1	0	1	1	0	0
1	1	0	1	1	0	0
1	0	0	1	1	0	0
1	0	1	1	1	0	0
1	1	1	1	1	1	1

TABLE 4.11
Truth Table for Distributive Property

INPUT			LOGIC			
X	Y	Z	X+(Y•Z)	(X+Y)(X+Z)	X•(Y+Z)	(X•Y)+(X•Z)
0	0	0	0	0	0	0
0	0	1	0	0	0	0
0	1	1	1	1	0	0
0	1	0	0	0	0	0
1	1	0	1	1	1	1
1	0	0	1	1	0	0
1	0	1	1	1	1	1
1	1	1	1	1	1	1

The product of an element times the sum of two or more elements is the same as the sum of the products of each of the elements times that element (Table 4.11).

Absorbtive

$$X + XY = X$$
$$X(X + Y) = X$$

(4.15)

The sum of an element and the product of that element with another element is equal to that element (Table 4.12).

Tautology

$$X + X = X$$
$$XX = X$$

(4.16)

TABLE 4.12
Truth Table for Absorbtive Property

INPUT		LOGIC	
X	Y	X+(X•Y)	X•(Y+X)
0	0	0	0
1	0	1	1
0	1	0	0
1	1	1	1

TABLE 4.13
Truth Table for Tautology Property

INPUT	LOGIC	
X	X+X	X•X
0	0	0
1	1	1

TABLE 4.14
Truth Table for Universal Class Property

INPUT		LOGIC	
X	1	X•1	X+1
0	1	0	1
1	1	1	1

The sum of two or more of the same element or the product of two or more of the same element is equal to that element (Table 4.13).

Universe Class

$$X \bullet 1 = X$$
$$X + 1 = 1$$

(4.17)

The product of an element and 1 is equal to that element. The sum of an element and 1 is equal to 1 (Table 4.14).

Null Class

$$X \bullet 0 = 0$$
$$X + 0 = X$$

(4.18)

TABLE 4.15
Truth Table for Null Class Property

INPUT		LOGIC	
X	1	X•0	X+0
0	1	0	0
1	1	0	1

TABLE 4.16
Truth Table for Complementation Property

INPUT		LOGIC	
X	X′	X•X′	X+X′
0	1	0	1
1	0	0	1

TABLE 4.17
Truth Table for Double Negation Property

X	X′	X″
0	1	0
1	0	1

The product of an element and 0 is equal to 0. The sum of an element and 0 is that element (Table 4.15).

Complementation

$$X + X' = 1$$
$$XX' = 0$$

(4.19)

The sum of an element and that element negated is equal to 1. The product of an element and that element negated is equal to 0 (Table 4.16).

Double Negation

$$(X')' = X'' = X$$

(4.20)

An element which has been negated twice is equal to that element (Table 4.17).

TABLE 4.18
Truth Table for Contraposition Property

X	Y	X'	Y'
0	1	1	0
1	0	0	1

TABLE 4.19
Truth Table for Expansion Property

INPUT		LOGIC	
X	Y	(X•Y)+(X•Y')	(X+Y)•(X+Y')
0	0	0	0
1	0	1	1
0	1	0	0
1	1	1	1

Contraposition

$$X = Y' \Rightarrow Y = X'' \tag{4.21}$$

The product of an element times a second element negated yields a product of the second element times the first element double negated (Table 4.18).

Expansion

$$XY + XY' = X$$
$$(X + Y)(X + Y') = X \tag{4.22}$$

The product of two elements or the product of an element and the second element negated is equal to the state of the first element. One could deduce this result by simply factoring out the first element (Table 4.19).

DeMorgan's Law

$$\overline{(X + Y)} = X'Y'$$
$$\overline{XY} = X' + Y' \tag{4.23}$$

This property is useful in changing from one logic function to another. Many times application of DeMorgan's law will allow resulting functions to be combined resulting in simplification of the logic (Table 4.20).

TABLE 4.20
Truth Table for DeMorgan's Property

INPUT		LOGIC			
X	Y	$\overline{X+Y}$	$\overline{Y \cdot X}$	X'•Y'	Y'+X'
0	0	1	1	1	1
1	0	0	1	0	1
0	1	0	1	0	1
1	1	0	0	0	0

TABLE 4.21
Truth Table for Reflection Property

INPUT		LOGIC			
X	Y	X+X'•Y	X•X'+Y	X+Y	X•Y
0	0	0	0	0	0
1	0	1	0	1	0
0	1	1	0	1	0
1	1	1	1	1	1

Reflection

$$X + X'Y = X + Y$$
$$X(X' + Y) = XY$$
(4.24)

Reflection allows sums and products containing both states to be simplified to a basic function (Table 4.21).

Transition

$$XY + YZ + X'Z = XY + X'Z$$
$$(X + Y)(Y + Z)(X' + Z) = (X + Y)(X' + Z)$$
(4.25)

Transition states that three or more functions can be reduced to two functions with the same output results by combining the combinations of outputs of the functions (Table 4.22).

Transposition

$$XY + X'Z = (X + Z)(X' + Y)$$
$$(X + Y)(X' + Z) = XZ + X'Y$$
(4.26)

See Table 4.23, Truth Table for Transposition Property.

TABLE 4.22
Truth Table for Transition Property

INPUT			LOGIC			
X	Y	Z	(XY)+(X′Z)+(YZ)	(XY)+(X′Z)	(X+Y)(X′+Z)(Y+Z)	(X+Y)(X′+Z)
0	0	0	0	0	0	0
0	0	1	1	1	0	0
0	1	1	1	1	1	1
0	1	0	0	0	1	1
1	1	0	1	1	0	0
1	1	1	1	1	1	1
1	0	1	0	0	1	1
1	0	0	0	0	0	0

TABLE 4.23
Truth Table for Transposition Property

INPUT			LOGIC					
X	Y	Z	X•Y	X′•Z	X+Z	X′+Y	XY+X′Z	(X+Z)•(X′+Y)
0	0	0	0	0	0	1	0	0
0	0	1	0	1	1	1	1	1
0	1	1	0	1	1	1	1	1
0	1	0	0	0	0	1	0	0
1	1	0	1	0	1	1	1	1
1	0	0	0	0	1	0	0	0
1	0	1	0	0	1	0	0	0
1	1	1	1	0	1	1	1	1

Boolean relationships can contain both literals and terms. Terms are combinations or groups of literals. Terms can be product or sums. A product term is a group of two or more inputs which are connected by AND. A SUM term is a group of two or more inputs connected by OR.

A conjunctive expression contains a single sum term or a product of sum terms. A disjunctive expression contains a single product term or a sum of product terms.

Conjunctive expressions:

$$X + Y = U$$
$$(X + Y)(A + B) = U \qquad (4.27)$$
$$A(B + C)(X + Y + Z) = U$$

Disjunctive expressions:

$$XY = U$$
$$XY + AB = U \qquad (4.28)$$
$$A + BC + XYZ = U$$

4.9 UNCOMPLEMENTATION

Since Boolean statements are always with dealing two states, two operations, etc. it can be shown that if all the states and all the operations are changed to the other state and other operation, the expression is not changed in value. This process is called complementation. An example of complementation is shown in:

$$XY' + XYZ' = (X' + Y)(X' + Y' + Z) \qquad (4.29)$$

Complemented Boolean statements can be cumbersome and difficult to analyze especially where NAND and NOR functions are combined with AND and OR. DeMorgan's law can be applied to simplify such statements using a process known as uncomplementation. Many times uncomplementation can be used to simplify cumbersome and complex statements.

4.10 SUMMARY

The relationship between binary and real world conditions can be used to simplify complex situations and allow accurate prediction of outcomes. The use of the traditional terms "normally open" and "normally closed" should be avoided because of the ambiguity that exists between electrical and fluid power use of the terms. ON/OFF, 0 state/1 state, passing/nonpassing are more useful terms. Boolean algebra is a powerful tool to allow concise modeling of logic combinations. The fundamentals of Boolean algebra should be thoroughly understood before proceeding to application of logic to real world applications. Boolean postulates are useful to develop an appreciation of the behavior of binary logic combinations.

CHAPTER 4 REVIEW QUESTIONS AND PROBLEMS

1. Show that X'+X=X and XCX=X.
2. Show that X+X'=1 and XCX'=0.
3. Prove that X+X'Y=X+Y for all conditions of X and Y.
4. Prove that for any condition of X, Y, Z the following expressions are equal:
   (X+Y) (X'+Z) (Y+Z)
   (XZ) + (X'Y) + (YZ)
   (X+Y)(X'+Z)
   (XZ) + (X'Y)
5. Reduce XY + X' + Y' to a single output value.

6. Simplify $(X+Y+Z)(XY+X'Z)'$ (Reduce to the smallest number of terms).
7. Simplify $(X'Y+XYZ'+XY'Z+X'Y'Z't+t)$.
8. Simplify $(X+Y')(Y+Z')(Z+X')(X'+Y')$.
9. Simplify $XY'+XZ+XY$.
10. Simplify $XYZ+(X+Y)(X+Z)$.
11. Simplify $(X+Y)(X+Y')(X'+Z)$.
12. Simplify $X'YZ+XY'Z'+X'Y'Z+X'YZ'+XY'Z+X'Y'Z'$.
13. Write the simplest expression for each of the functions f1, f2, f3, f4 for the conditions tabulated below.

TABLE 4.24
Conditions for Problem 13

Conditions	X	Y	Z	f1	f2	f3	f4
1	1	1	1	1	0	0	1
2	1	1	0	0	1	1	1
3	1	0	1	1	0	0	1
4	1	0	0	1	0	1	0
5	0	1	1	1	0	1	1
6	0	1	0	1	0	1	1
7	0	0	1	0	1	0	1
8	0	0	0	1	0	0	0

14. If p is ON, q is OFF, and r is OFF, determine the output state of the following:
   a. p'
   b. pq
   c. $p+q$
   d. $(p+q)+r$
   e. $p'+(q+r)$
   f. $p'+(q+r)'$
   g. $pq+qr$
   h. $pq+p'q'$
   i. $(pq+p'q')+(p'q+pq')$
   j. $(p'q')[(p+r')(q+r)]$
15. Determine the output state of the $f(p,q,r)$ in problem 14 if p, q, r are all ON.

5 MPL Concepts and Components

MPL refers to moving-parts logic which uses pneumatic devices such as valves that have moving parts. Most air logic circuits use MPL. A wide variety of pneumatic components are available which can be used for air logic control. Some devices are miniaturized to allow high logic density. Many air logic circuits are used to control fluid power devices and are often interfaced with electrical controls to operate electrical power equipment. Because of this coincidence, familiarization with power components as well as control components is necessary for correct application and analysis of logic circuits.

5.1 CONCEPTS

Although any valve can be used as an MPL element, devices designed specifically for MPL are miniaturized valves which can be interconnected through specially designed manifolds. Miniaturization allows greater logic density, and manifolding reduces the amount of external piping required. Concepts commonly exploited for MPL logic devices include:

1. Spool
2. Diaphragm
3. Poppet
4. Floating spool

5.2 SPOOL VALVE CONSTRUCTION

The most common construction for an MPL device is the spool valve. Figure 5.1 shows the construction of a basic spool device. Note that one can easily configure the spool to switch multiple port arrangements to produce most logic functions. Since the pressure area is the same for all ports, the valve is balanced allowing ports to be used as inputs or outputs providing for variations in function simply by connecting to the appropriate ports. Also since the spool is balanced, very low operating pressures can be used to operate these valves. The spool construction allows for miniaturization resulting in relatively high logic function density.

5.3 DIAPHRAGM VALVE CONSTRUCTION

The diaphragm construction produces an unbalanced valve, so ports are dedicated and the valve must be designed for a specific logic function. Advantages of this design are simplicity of construction and excellent sealability due to the rubber diaphragm being pressurized against a metal seat (Figure 5.2).

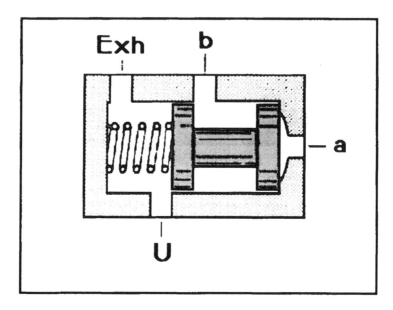

FIGURE 5.1   Cross section showing spool valve construction

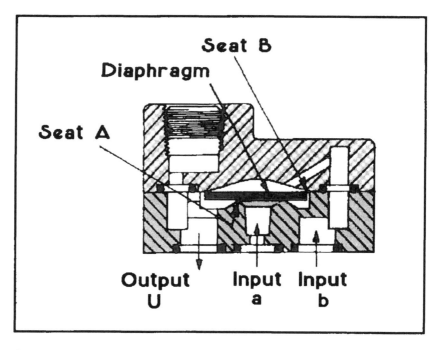

FIGURE 5.2   Cross section of an MPL diaphragm valve

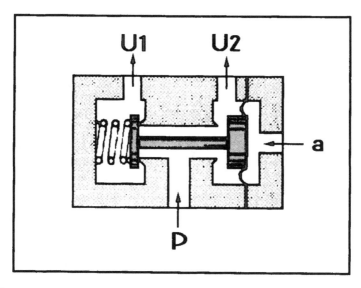

FIGURE 5.3   Cross section of a poppet construction MPL device

5.4   POPPET VALVE CONSTRUCTION

The poppet construction combines the spool with the diaphragm resulting in a fast response construction with a short operating stroke. This design however is limited to three port designs and is dedicated to a particular function limiting its versatility (Figure 5.3).

5.5   FLOATING SPOOL CONSTRUCTION

The floating spool construction is limited to a few logic functions and is normally a passive device. The major advantage of the floating spool design is its simplicity (Figure 5.4).

5.6   MPL CLASSES

Two classes of MPL devices are used:

1. Low pressure
2. High pressure

Low pressure MPL devices typically operate with pressures between 1–30 psig and have internal flow passages between .03 and .1 inches in diameter. Because of the small flow passages, air must be fine filtered and free from particulates, water, and oil. Standard regulators are adequate for pressure control to these devices. Low pressure valves have small flow capacities and may be used for logic functions or operators for high pressure valves.

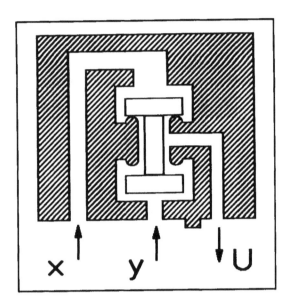

FIGURE 5.4 Floating spool construction MPL device

High pressure MPL devices typically operate with pressures between 30 and 125 psig and have internal flow passages between .1 and .2 inches in diameter. Coarse filtering of air for these devices is usually adequate. Lubrication in the air is usually required. High pressure devices can be used for logic functions and to operate power equipment.

MPL devices can also be classified as active or passive. Active devices require a separate supply pressure in addition to signal pressures, and the output is a result of switching the supply pressure on or off. Passive devices have no separate supply pressure and the output from the device is the switched signal pressure.

The signal pressures of most MPL devices are isolated from the environment and can be treated as static pressure. Since little or no flow takes place in the signal lines, inlet and outlet pressures can be considered nearly constant in short lines. Signal pressures can therefore be treated as digital signals, either pressure or no pressure (pressure ON or pressure OFF).

To avoid ambiguity, two states or conditions must be defined for MPL devices. The 1 state is obtained by the presence of pressure and the 0 state is obtained by exhausting pressure to atmosphere. Signal and output states may be the same and are referred to as normally closed. Valves having differing signal and output states are referred to as normally open (Table 5.1).

5.7 TYPICAL MPL VALVE OPERATION

Figures 5.5 and 5.6 illustrate a normally closed spool valve. Note the position of the spool with no signal pressure on the end of the spool blocks pressure from the supply to the output. Signal pressure applied to the end of the spool overcomes the

TABLE 5.1
Valve States

	Signal	Output
Normally Closed	No Pressure	No Flow
	0 State	0 State
	Pressure On	Flow On
	1 State	1 State
Normally Open	No Pressure	Flow On
	0 State	1 State
	Pressure On	No Flow
	1 State	0 State

spring force on the other end of the spool shifting the spool and allowing the supply air to flow to the output.

The normally open configuration illustrated in Figures 5.7 and 5.8 shows the position of the spool with no signal pressure with the supply and output ports connected allowing flow through the valve. When signal pressure is applied to the end of the spool the spring force is overcome shifting the spool to block the flow path isolating the output from the supply.

Many MPL devices have dual output ports, so NO or NC outputs may be selected. Also it is common practice to provide exhaust ports which vent the output to atmosphere when the output is closed. Spool valves can be made with several ports and flow paths which allow combinations of logic functions in the same valve. Since operation of the spool involves overcoming the spring pressure by an applied signal pressure, a delay exists during signal application. This delay between signal application and output response can be minimized by making the spool low friction and by using low spring rate springs. Where response time is critical, it is important to consider the valves time constant. Most suppliers provide this data with the valve's specs.

The poppet valve has much faster response time because of its large ratio between diaphragm area and the seat area. The fast action of the poppet is useful in converting slow changing analog signals to digital on/off output. Typically response times for MPL devices exceed 10ms as compared to electronic devices which operate below 1 ms response time. The slower response time can be an advantage in many industrial control applications, especially operations which mimic human movements.

Another advantage of MPL devices is their ability to deliver relatively large amounts of power. MPL logic devices typically deliver as much as 1000 times the power as their electronic counterparts. High power outputs are very advantageous in many automation applications. Power consumption of pneumatic devices is directly proportional to the cycling rate. The device consumes power only when switching. Other types of devices such as solid state electronic logic devices, consume power during all modes of their operation. This characteristic can be advantageous where efficiency and low power loss is critical.

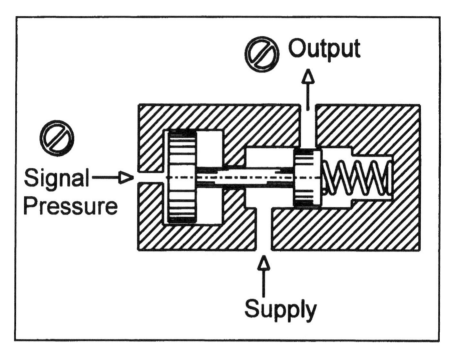

FIGURE 5.5 Normally closed MPL device with no signal applied

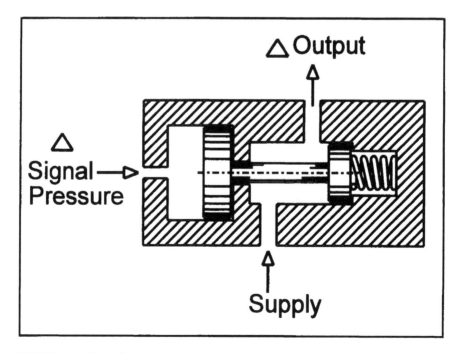

FIGURE 5.6 Normally closed MPL device with signal applied

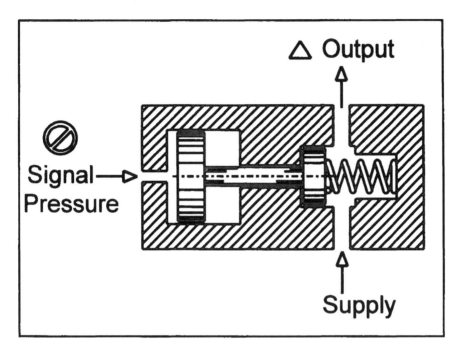

FIGURE 5.7   Normally open MPL device with no signal applied

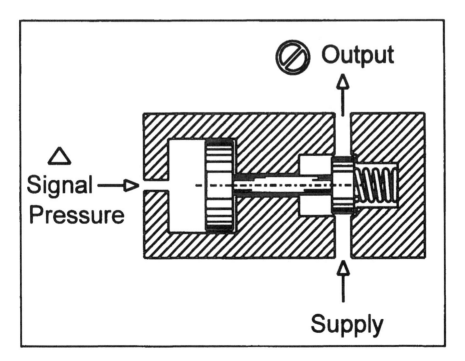

FIGURE 5.8   Normally open MPL device with signal applied

TABLE 5.2
Characteristics of MPL Devices

Slow response time
Large power delivery
Relatively efficient
Multiple functions
Multiple outputs
Reliable and rugged
Various inputs possible
Can be active or passive
Can be manifolded easily
Operate at high pressures
Miniaturization possible
Most logic functions possible
Explosion proof
Produce volatile memory
Can control, amplify, interface

MPL devices are available with multiple port arrangements which function as a variety of logic elements. Logic elements are defined as devices capable of producing a 0 or 1 state output condition based on its input conditions. The basic functions are described in Chapter 7, Logic Elements.

A major disadvantage of MPL devices is their size. Despite a great deal of miniaturization, MPL logic elements occupy a great deal of volume, therefore high density logic systems are not possible with ALC. Future developments in nanotechnology may allow high density air logic approaching the densities of electronic elements.

Inputs to MPL logic devices can be:

- Pneumatic
- Hydraulic
- Electric
- Mechanical

Type of input is dependent upon the application of the device and how it must be interfaced with its immediate world. Most manufacturers design their valves so operators are interchangeable and electrical, pneumatic, or mechanical operators may be selected for any device.

Since pressure pulses in a compressible fluid are attenuated as they flow along a conduit, lengths of effective transmission are limited to several feet.

5.8   SUMMARY

Many designs of MPL devices or valves suitable for MPL exist. Selection of the proper device is usually a matter of compromise between function, size constraints, and cost. The simplest construction is usually the best. Miniaturization has provided an even greater variety of devices which combine functions or include multiple functions in one package. Novel port arrangements have been developed to allow systematic connecting and manifolding of devices. The primary benefits of MPL are low cost, ruggedness, high reliability, slow response time, and delivery of a great deal of power. Although many designs of valves have been developed for MPL applications, any valve can be used which has the proper operational logic (Table 5.2).

CHAPTER 5 REVIEW QUESTIONS AND PROBLEMS

1. Name two classes of MPL devices.
2. Why are most pneumatic systems limited to 125 psi?
3. Describe 1 state and 0 state.
4. Define signal pressure.
5. What logic function is produced with a single normally closed valve?
6. What logic function is produced with a single normally opened valve?
7. How can response time be minimized with an MPL device?
8. Define response time.
9. What is a typical response time for an MPL device?
10. How will response time of a device affect a control system?
11. Name and describe 4 types of inputs (signals) to MPL devices.
12. Why is signal line length critical to an MPL system?
13. Describe a poppet type MPL device.
14. Describe a spool type MPL device.
15. Describe a floating spool MPL device.
16. Describe a diaphragm MPL device.
17. Describe the major advantages of MPL devices.
18. Describe the major disadvantages of MPL devices.

6 NonMPL Concepts and Components

NonMPL systems involve the use of fluidic concepts. Fluidic devices have no moving parts and operate based on dynamic fluid properties and phenomenon. They can be used for sensing, logic, memory, and timing. Fluidic devices operate on pressure inputs usually below 10 psig and can provide analog or digital outputs. Their application suitability depends upon three factors: operating pressure, flow during operation which is a function of operating pressure and passage diameter, and power which is a result of both the operating pressure and flow. Vent ports and flow restrictors are used in some devices to maintain required internal pressure balance and to allow the operating phenomenon to exist.

6.1 FLUIDIC CONCEPTS

Two fluidic concepts used for nonMPL devices are jet deflection and jet destruction. Fluidic devices can also be either active or passive. Active devices require a separate supply pressure, and signal pressures divert the supply to the appropriate output. Active devices also act to amplify the input signals, and so may have an analog as well as a discrete output. Passive devices do not have separate supply ports, and they do not amplify the input. They use the input signals to produce their outputs by simply deflecting the input signals to the corresponding outputs (Figures 6.1 and 6.2).

6.2 JET DEFLECTION

Jet deflection devices operate by exploiting phenomenon which divert the supply or signal pressure to one output or another. A jet is a stream of fluid moving in one direction. Jets can be either laminar (low velocity) or turbulent (high velocity); usually, the turbulent or high speed jet is produced in most fluidic devices. The jet is maintained and is not destroyed by the diversion. If a separate supply pressure is used, the device is said to be active. If the signal pressures are used to produce the output, the device is considered to be passive. Jet deflection devices operate in a pressure range of between 2 and 10 psig and have flow passages between .01 and .03 inches in diameter. They require clean, dry air for reliable operation. Methods used for jet deflection are jet interaction, vortex feedback, wall attachment, and wall reflection (Table 6.1).

Jet Interaction

This technique uses a signal input normal to the jet. Signal pressure deflects the jet by the impending velocity at right angles to it, causing a resultant velocity deflected

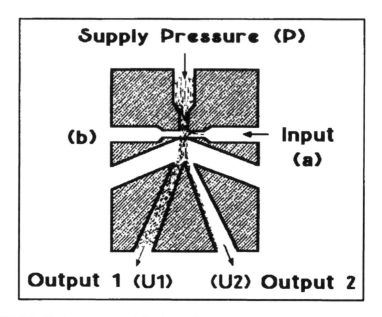

FIGURE 6.1   Typical active jet deflection device

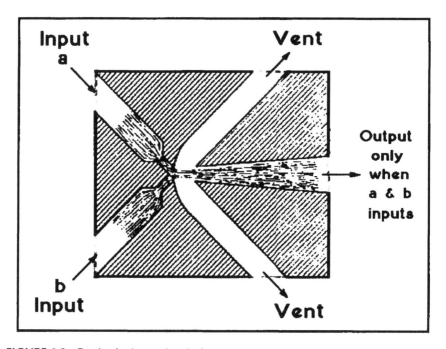

FIGURE 6.2   Passive jet interaction device

TABLE 6.1
Jet Deflection Devices

Concept	Active	Passive
Jet Interaction		Inclusive OR
		AND
		Exclusive OR
Wall Attachment	YES	
	NOT	
Vortex Feedback	OR	
	AND	
Wall Reflection	Memory	
	Amplifier	

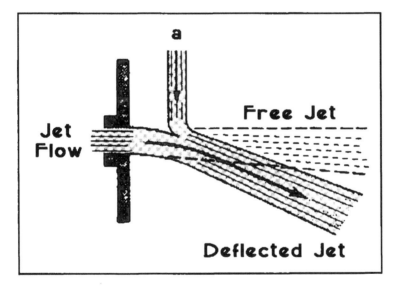

FIGURE 6.3   Deflection by jet interaction

from the free jet. The amount of deflection depends upon the pressure of the signal, so the output can be modulated by the signal providing an analog output (Figure 6.3).

Vortex Feedback

This device includes a vortexing chamber which provides a deflection stream taken off the main jet that provides a deflection pressure to the main jet. The vortexing chamber is a means of amplifying a signal pressure to deflect the supply jet and will hold the jet in that position even after the signal pressure is removed. Application of a signal pressure on the opposite side of the jet is required to switch it back to the other output port (Figure 6.4).

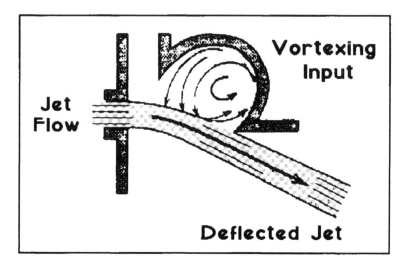

FIGURE 6.4   Deflection by input vortex

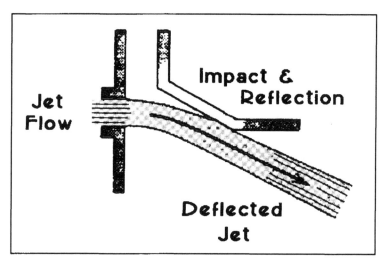

FIGURE 6.5   Jet deflection by reflection

Wall Reflection

Wall reflection devices use chambers on each side of the supply stream which divert
the signal pressure in a manner that attracts the supply stream, switching the supply
jet to the side where the signal pressure is applied. Note that all other methods switch
the supply to the opposite side from which the signal is applied (Figure 6.5).

Wall Attachment

The wall attachment device operates using a phenomenon known as the Coanda
Effect. A signal pressure will deflect the supply pressure of this device to one of

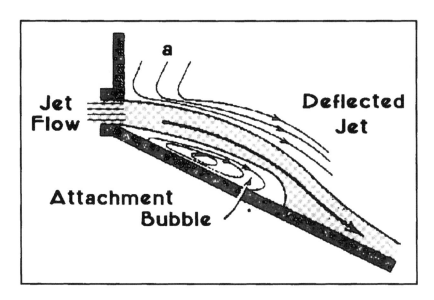

FIGURE 6.6   Wall attachment of a deflected jet

two output ports. Because of wall attachment when the signal pressure is removed, the supply will remain flowing to the port to which it was deflected. Signal pressure at the opposite signal port will deflect the supply to the other output where it will remain until a signal pressure is applied to the opposite signal port. This phenomenon provides a memory to the device which remembers which port the signal pressure was applied to last (Figure 6.6).

6.3   JET DESTRUCTION

Jet destruction devices operate by impinging a signal pressure on a jet stream and destroying it (causing turbulence) to effectively shut off the output. These devices operate with pressures under .5 psig supply and as low as .02 psig signal. Passages in the devices are very small compared to the jet deflection devices, and therefore require extremely clean, dry air for reliable operation. Methods used for jet destruction devices include laminar/turbulent flow devices and impact modulators (Table 6.2).

Laminar/Turbulent Flow Devices (Turbulence Amplifiers)

A very low supply pressure creates a laminar flow between the outlet and inlet which are separated by a gap at atmospheric pressure. Flow reaches the outlet exiting the output port. At the gap, a signal pressure may be applied which mixes with the laminar flow diverting it from the output. The area where this occurs is referred to as the interaction chamber. This interaction destroys the flow to the output and vents it to atmosphere, shutting off the output. Typical inlet pressures are less than $\frac{1}{2}$ psig, with output pressures around .2 psi. Typical signal pressure to completely shut off are less than .1 psig. The turbulence amplifier may be designed with several signal

TABLE 6.2
Jet Destruction Devices

Concept	Devices
Laminar/Turbulent	OR
	YES
	NOT
	AND
	Memory
	Amplifiers
Impact Modulator	OR
	YES
	NOT
	AND
	Memory
	Amplifiers

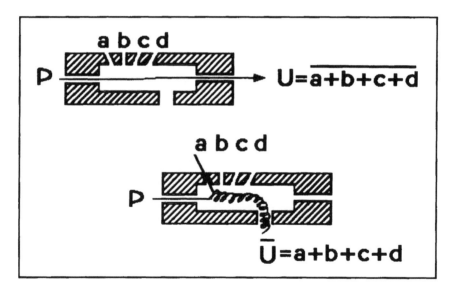

FIGURE 6.7   Jet destruction concept

inputs resulting in a NOR gate which shuts off the output with any combination of signal inputs (Figure 6.7).

Impact Modulators

A typical impact modulator has two supplies flowing toward each other but are separated by an orifice. The supply on one side is slightly higher in pressure than the other, keeping the second supply from exiting through the orifice. In this state,

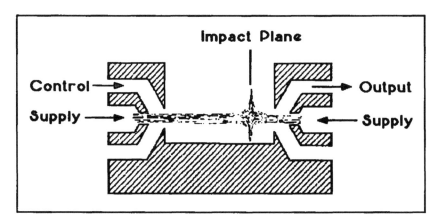

FIGURE 6.8   Impact modulator concept

the supply exits through the outlet creating an output signal. When a signal is applied on the side of the stronger supply, this signal opposes the supply and allows the opposite supply to exit through the orifice resulting in shutting off the output. Impact modulators may be designed to operate with higher pressures since the supply flow can be either laminar or turbulent. Typical pressures can be between .5 and 5 psig with signal pressures as low as .01 psig. Several signal ports may be used to create a NOR device using impact modulation (Figure 6.8).

6.4   FLUIDIC SENSORS

Fluidic sensors are usually used as proximity devices that can detect the presence of an object and provide an output signal announcing its presence. Most fluidic sensors act as bleed valves so air consumption must be minimized. To reduce air consumption, the supply port is kept as small as possible, and nozzles producing the jets are designed between .004 and .010 diameter to produce an acceptable flow. Fluidic sensors are convenient input devices to air logic systems since they can be coupled directly to ALC devices with no interface of any kind. Sensor outputs usually need amplification by an ALC device when used to operate a pneumatic (or hydraulic) power device. Fluidic amplifiers are often used for these applications (Table 6.3).

TABLE 6.3
Fluidic Sensor Concepts

Concept	Use
Back Pressure Switch	Contact switch and proximity up to .02" away
Cone Jet	Proximity sensing up to .2" away from sensor
Interruptible Jet	Proximity up to 4" away from sensor
Ultrasonic Sensor	Proximity up to 5 feet from sensor

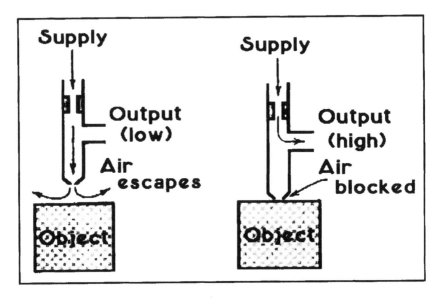

FIGURE 6.9   Back pressure sensor operation

Back Pressure Switching

The back pressure sensor is a nozzle combined with an OR/NOR gate. When no objects are present in front of the nozzle, the flow through the nozzle drops the supply pressure sufficiently to provide a very low pressure to the gate. When an obstruction covers the nozzle, a back pressure results due to retardation of the flow through the nozzle, and high pressure is developed at the gate switching it to its opposite state. This switched output can be used to control another logic device or operate a power valve or switch (Figure 6.9).

Cone Jet

This sensor has an annular nozzle surrounding a sensing orifice which diverts flow to the output. High pressure flow exits the annulus leaving a low pressure at the output orifice. When an object comes close to the nozzle part of the output, the jet is reflected back off the object to the orifice increasing the output pressure. The cone jet can be used to sense objects further away than the back pressure sensor (Figure 6.10).

Interruptible Jet

The interruptible jet sensor uses an air stream which is transmitted across a gap to the output. When an object interrupts the flow across the gap, the output pressure is decreased sufficiently to switch a gate which is incorporated in the device (Figure 6.11).

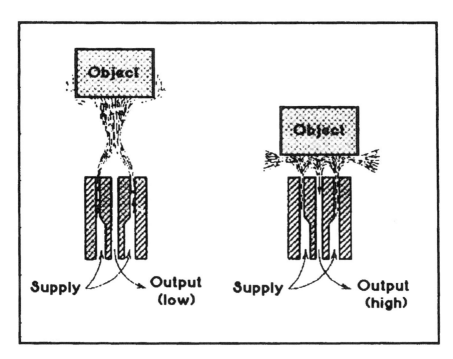

FIGURE 6.10   Cone jet sensor operation

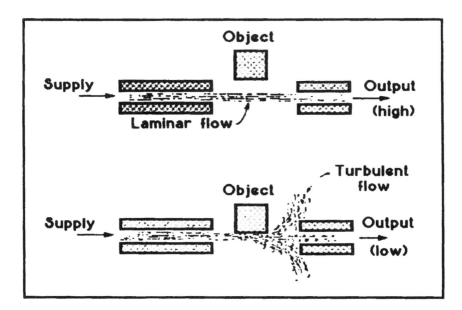

FIGURE 6.11   Interruptible jet sensor operation

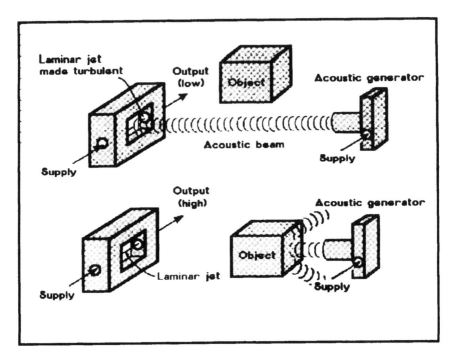

FIGURE 6.12   Acoustical sensor concept

Ultrasonic (Acoustic) Sensors

This sensor uses sound waves to disrupt a jet stream, keeping the output shut off. The sound waves are produced by a transmitter referred to as a "whistle." The sound waves are inaudible and are usually around 50 kHz. The laminar jet in the receiver is destroyed by the emitted tone acting as an input signal. When the sound is blocked by the presence of an object between the transmitter and the receiver, the jet is restored. Ultrasonic sensors have a range up to 5 feet in the direct mode and up to 1 foot in the reflected mode (Figure 6.12).

6.5   DEVICE PACKAGES

Fluidic devices are much larger than their electronic counterparts but usually smaller than MPL devices. Typical dimensions for an active fluidic device package are 1" × 1.5" × 1/4" thick. Fittings are usually sized for .170 I.D. plastic tubing. Pressure and signal ports are usually on the back side allowing manifold coupling for maximum logic density although some model have ports on the sides (Figure 6.13).

6.6   SUMMARY

NonMPL devices can be used for a variety of logic functions. Their primary advantage is they operate with no moving parts so they do not encounter wear and

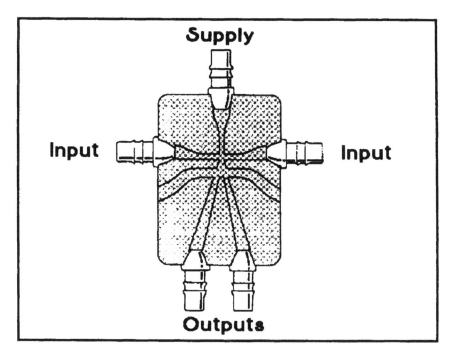

FIGURE 6.13 Typical active fluidic device package

fatigue. Their main disadvantage is their small port sizes require extremely clean air for reliable service. NonMPL devices are especially useful in many sensor applications.

CHAPTER 6 REVIEW QUESTIONS

1. Describe the difference between active and passive devices.
2. Define a jet.
3. Describe jet deflection and the types of devices using jet deflection.
4. Describe jet destruction and the types of devices using jet destruction.
5. Describe the major types of fluidic sensors.
6. Describe some applications for fluidics. What are the benefits of fluidics in these applications?
7. What are some disadvantages of fluidic devices?
8. What are the major advantages of fluidic devices?

7 Logic Elements

Any air logic control network is made up of basic logic elements. Logic elements perform some basic logic function. These functions include gates, memory, and timers. Many components are constructed to perform one or more logic functions; however, in some cases components need to be combined to perform a logic function. This chapter presents the basic logic elements in relay logic, MPL, nonMPL, and illustrates each function with its Boolean equation and truth table. A combination of valves or devices that perform a specific logical operation is referred to as a gate circuit or simply a gate. The following discussion of gates uses 1 or 2 valve circuits or signal inputs to simplify the discussion. Realize that the gate could have several inputs and multiple outputs. Gates can function as memories to retain the state of the input even after the input signal is removed. Provide a means of remembering the status of the input at a point in time. Timers allow operation to occur for a specified duration or be delayed by some predetermined time.

7.1 YES

The YES function has a single input (X) and a single output (U) with a 0 state (closed) device. With no input into the device, there is no output. When an input is applied, the device changes state to 1 (open) and there is flow through the device (output). The Boolean expression for this function is $U = X$. See Table 7.1, Figure 7.1, Figure 7.2, and Figure 7.3.

7.2 NOT

The NOT gate uses a normally open valve (1 state). When no input is on the valve, there is an output (1 state). With an input on the valve, there in no output (0 state). The Boolean expression for the NOT function is $U = X'$. See Table 7.2, Figure 7.4, Figure 7.5, and Figure 7.6.

7.3 AND

The AND function uses 0 state devices in series. Examples shown use 2 valves in series for simplification of truth tables. Actual applications of the AND circuit may include several devices in series. When the valves are in the normal (0 state) positions, there is no output. When any valve is changed to the 1 state, no output results. The output will not change to 1 state until all the valves in the series line are in the 1 state. The Boolean expression for the AND gate is $U = XY$. See Table 7.3, Figure 7.7, Figure 7.8, and Figure 7.9.

TABLE 7.1
Truth Table for YES Function

INPUT	OUTPUT
X	U
0	0
1	1

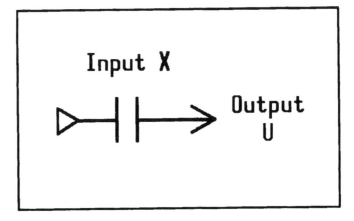

FIGURE 7.1 YES function using relay logic

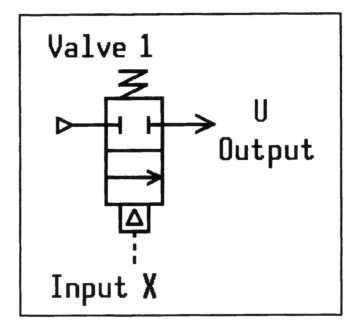

FIGURE 7.2 MPL YES function

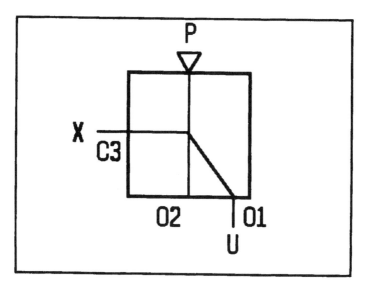

FIGURE 7.3 Fluidic YES function

TABLE 7.2
Truth Table for NOT Function

INPUT	OUTPUT
X	U
0	1
1	0

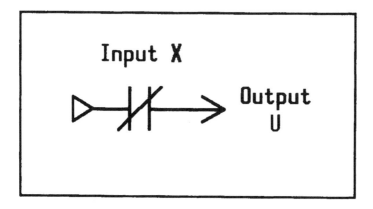

FIGURE 7.4 NOT function shown in relay logic

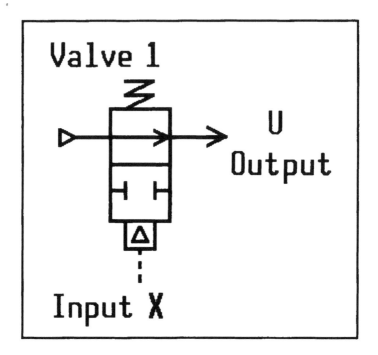

FIGURE 7.5 MPL NOT function

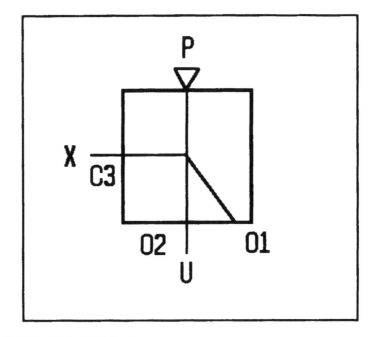

FIGURE 7.6 Fluidic NOT function

TABLE 7.3
Truth Table for AND Gate

INPUTS		OUTPUT
X	Y	U
0	0	0
1	0	0
0	1	0
1	1	1

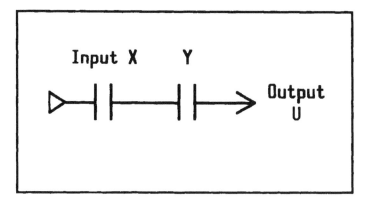

FIGURE 7.7 AND gate shown in relay logic

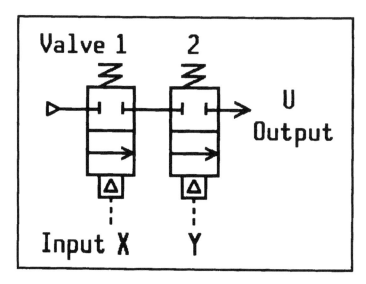

FIGURE 7.8 MPL AND gate

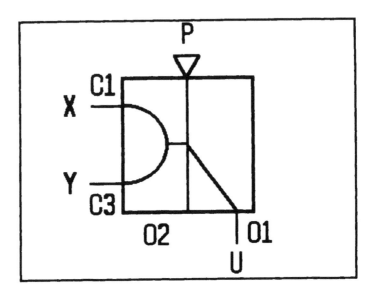

FIGURE 7.9 Fluidic AND gate

7.4 OR

The OR gate is formed by connecting 0 state valves in parallel. With no input on either valve, there is no output. If an input is applied to either or both valves, an output exists. The Boolean expression for the OR gate is U = X+Y. See Table 7.4, Figure 7.10, Figure 7.11, and Figure 7.12.

7.5 NAND

The negated AND or NAND gate has 1 state valves in parallel, resulting in an output until both valves are changed to 0 state. Note that the outputs in the truth table for the NAND gate are all opposite of the AND gate. The Boolean expression for the NAND gate is U = X'+Y' = (XY)'. See Table 7.5, Figure 7.13, Figure 7.14, and Figure 7.15.

TABLE 7.4
Truth Table for OR Gate

INPUTS		OUTPUT
X	Y	U
0	0	0
1	0	1
0	1	1
1	1	1

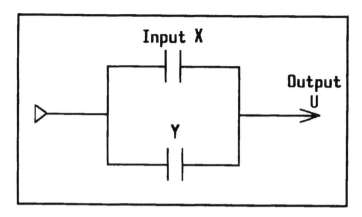

FIGURE 7.10   OR gate shown in relay logic

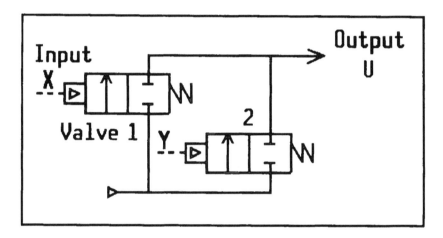

FIGURE 7.11   MPL OR gate

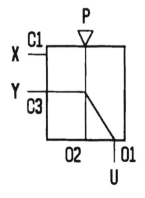

FIGURE 7.12   Fluidic OR gate

TABLE 7.5
Truth Table for NAND Gate

INPUTS		OUTPUT
X	Y	U
0	0	1
1	0	1
0	1	1
1	1	0

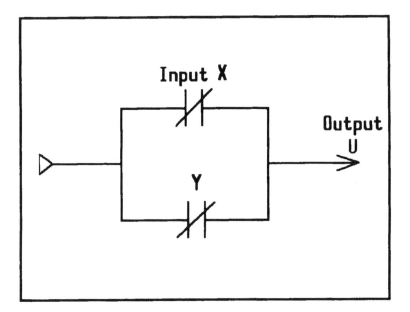

FIGURE 7.13   NAND gate shown with relay logic

7.6   NOR

The NOR gate has 1 state valves connected in series. The NOR gate has an output with no input. An input to either or both valves will change the valve state to 0 and stop and output. Note that the output in the truth table for the NOR gate is opposite of the OR gate. The Boolean expression for the NOR gate is $U = X'D' = (X+Y)'$. See Table 7.6, Figure 7.16, Figure 7.17, and Figure 7.18.

7.7   EXCLUSIVE OR

The Exclusive OR functions the same as the OR gate except output exists only when individual inputs are changed to 1 state. Combinations of inputs in the 1 state do not produce an output. The Boolean expression for this function is $U = X'D+XY'$. See Table 7.7, Figure 7.19, Figure 7.20, and Figure 7.21.

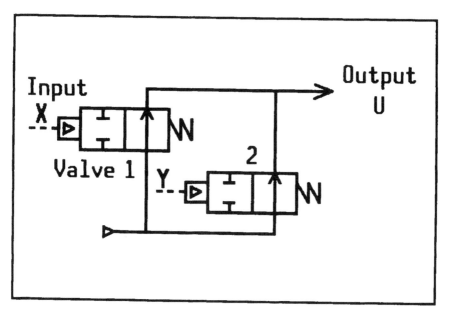

FIGURE 7.14   MPL NAND gate

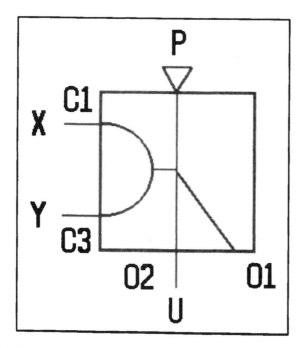

FIGURE 7.15   Fluidic NAND gate

TABLE 7.6
Truth Table for NOR Gate

INPUTS		OUTPUT
X	Y	U
0	0	1
1	0	0
0	1	0
1	1	0

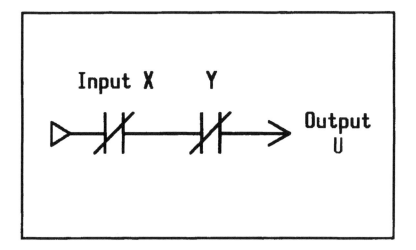

FIGURE 7.16   NOR gate shown with relay logic

7.8   COINCIDENCE

The Logic Coincidence produces an output only when all inputs are equal. The Boolean expression for this function is $U = XY+X'D'$. See Table 7.8, Figure 7.22, Figure 7.23, and Figure 7.24.

7.9   INHIBITOR

The Logic Inhibitor produces an output 1 state only when no signal is applied at X and a signal is applied at Y. The Boolean expression for the logic inhibitor is $U = X'D$. See Table 7.9, Figure 7.25, Figure 7.26, and Figure 7.27.

7.10   TIMERS

The Logic Timer is a one shot element. When input X changes to 1 state, the output U changes to 1 state but only after a predetermined period of time. The predetermined time returns to 0 state even though the input remains at 1 state. The time period at

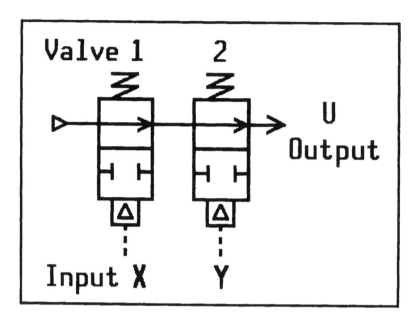

FIGURE 7.17   MPL NOR gate

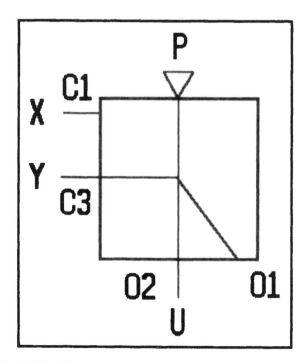

FIGURE 7.18   Fluidic NOR gate

TABLE 7.7
Truth Table for the Exclusive OR Function

INPUTS		OUTPUT
X	Y	U
0	0	0
1	0	1
0	1	1
1	1	0

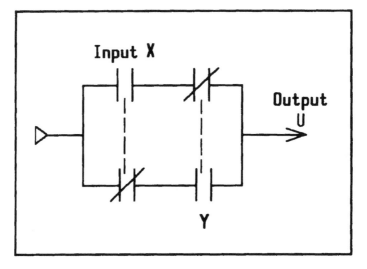

FIGURE 7.19   Relay logic for the exclusive OR function

1 state is determined by an orifice connecting the pressure source with the signal (Figures 7.28 and 7.29).

7.11   TIME DELAY

Timing IN

Timing IN delays the initiation of the state change by the time taken to fill the accumulator at a rate determined by the flow control the signal pressure must flow through. State change when the signal is removed is not timed because of free flow through the check valve in the flow control. Realize this same scheme could include the valve actuator for an MPL device as well as a fluidic device (Figure 7.30).

Timing OUT

State change when the signal is initiated is not timed because of free flow through the check valve in the flow control. Timing OUT delays the state change after the

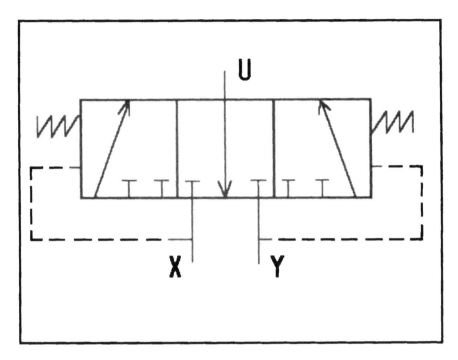

FIGURE 7.20   MPL exclusive OR using a 3 position 4 port valve (MPL)

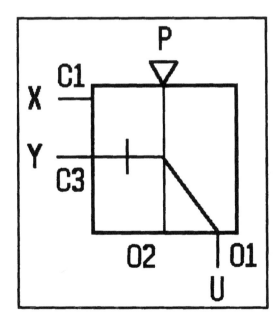

FIGURE 7.21   Fluidic exclusive OR function

TABLE 7.8
Truth Table for Logic Coincidence

INPUTS		OUTPUT
X	Y	U
0	0	1
1	0	0
0	1	0
1	1	1

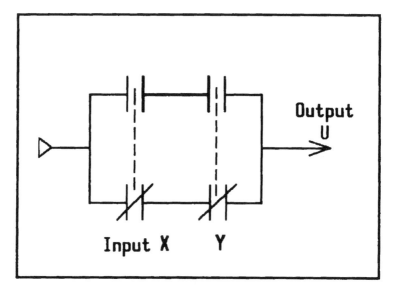

FIGURE 7.22   Relay logic for coincidence function

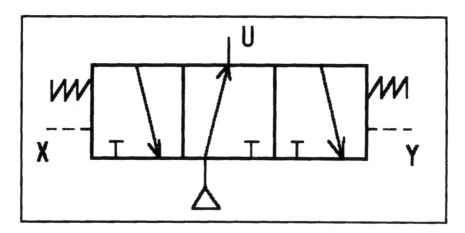

FIGURE 7.23   MPL coincidence function

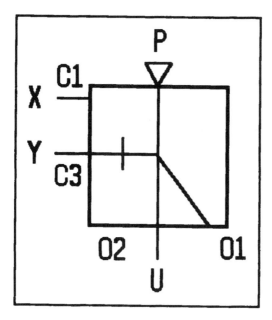

FIGURE 7.24 Fluidic coincidence function

TABLE 7.9
Truth Table for Logical Inhibitor

INPUTS		OUTPUT
X	Y	U
0	0	0
1	0	0
0	1	1
1	1	0

signal is removed. The delay time is dependent on the time required to empty the accumulator and the flow rate through the flow control valve. Realize this same scheme could include the valve actuator for an MPL device as well as a fluidic device (Figure 7.31).

7.12 FLIP-FLOPS

Basic Flip-Flop

The flip-flop is a memory device because it retains its state after the removal of the signal which produced that state. Referring to Figure 7.32, when a signal is applied to C1 or C3, the output pressure is switched to O1 and will remain at O1 after the signal pressure is removed. Signal pressure is required at C2 or C4 to switch the

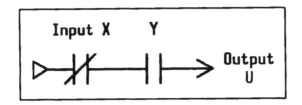

FIGURE 7.25   Relay logic inhibitor function

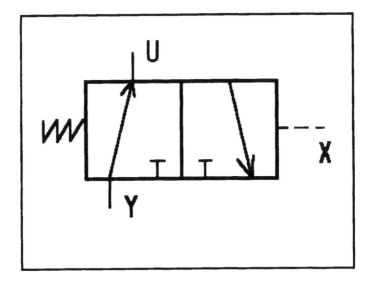

FIGURE 7.26   MPL inhibitor function

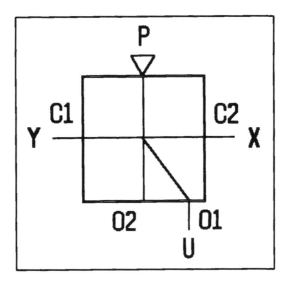

FIGURE 7.27   Fluidic inhibitor function

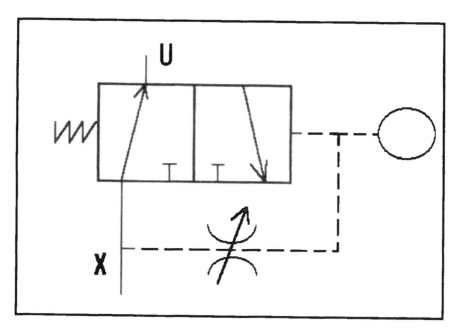

FIGURE 7.28   MPL timer

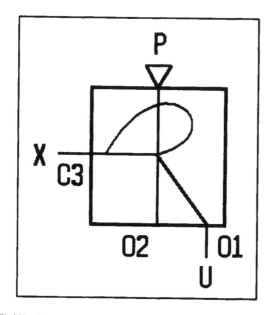

FIGURE 7.29   Fluidic timer

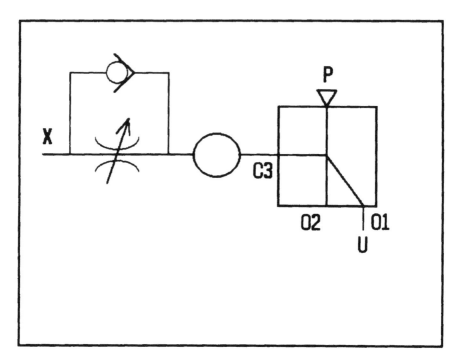

FIGURE 7.30   Fluidic time delay — Timing IN

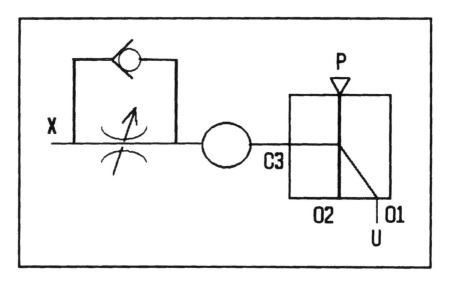

FIGURE 7.31   Fluidic time delay — Timing OUT

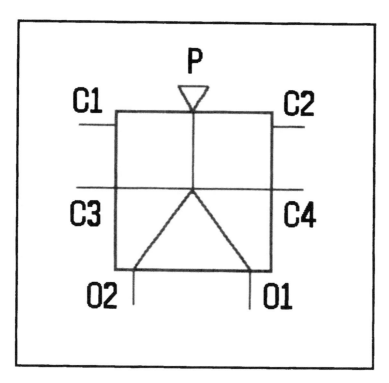

FIGURE 7.32   Basic fluidic flip-flop

TABLE 7.10
Truth Table for Basic Flip-Flop

INPUTS		OUTPUT	
C1	C2	O1	O2
1	0	1	0
0	0	1	0
0	1	0	1
0	0	0	1

output to O2 where it will remain until signal pressure is applied to C1 or C3 or the supply pressure is removed. At startup, with no signal applied, the output can be at either O1 or O2 (Table 7.10).

Preferenced Flip-Flop

Some applications require output at a specific port, usually O1, at startup with no signal applied. The preferenced flip-flop symbol shown in Figure 7.33 indicates output to O1 (marked with +) at startup with no signal applied. Otherwise the operation of the preferenced flip-flop is the same as the basic flip-flop (Table 7.11).

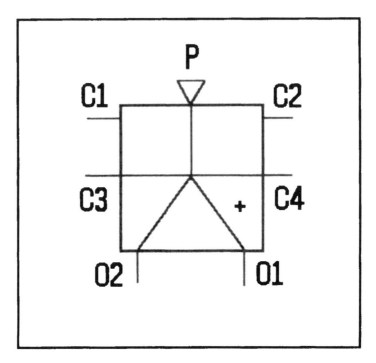

FIGURE 7.33   Fluidic preferenced flip-flop

TABLE 7.11
Truth Table for Preferenced Flip-Flop

INPUTS		OUTPUT	
C1	C2	O1	O2
0	0	1	0
0	1	0	1
0	0	0	1
1	0	1	0
0	0	1	0

SRT Flip-Flop

Operation of the SRT flip-flop is the same as the basic flip-flop except it has a trigger input which switches the output to the other port when a trigger pulse is applied. S and R designate SET and RESET and function identical to the signal input ports of the basic flip-flop. Signal pressure at S switches the output to O1 and signal pressure at R switches the output to O2. T switches the output to the other port regardless of the state of the output when the trigger was applied. The SRT is commonly used for binary counting circuits (see Table 7.12 and Figure 7.34).

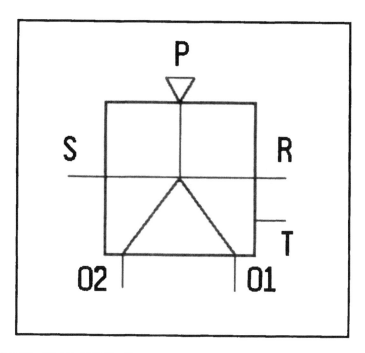

FIGURE 7.34   Fluidic SRT flip-flop

TABLE 7.12
Truth Table for SRT Flip-Flop

INPUTS			OUTPUT	
T	S	R	O1	O2
0	1	0	1	0
0	0	0	1	0
0	0	1	0	1
0	0	0	0	1
1	0	0	1	0
0	0	0	1	0
1	0	0	0	1
0	0	0	0	1

7.13   SUMMARY

MPL and fluidic devices are commercially available for the functions discussed in
this chapter and only basic functions have been included. Variations and combinations of these functions in common packages also are available and may be advantageous in specific applications. A great deal of miniaturization has evolved in the
MPL technologies which allows development of powerful control systems especially
with the use of multifunction valves.

CHAPTER 7 REVIEW QUESTIONS AND PROBLEMS

1. What type of logic function will operate (provide output to) a cylinder only when one of two safety switches provide pressure (signal) to the element? Show the truth table for this function.
2. Describe the logic operation of a Coincidence element. Show how this logic function could be produced using two three port valves.
3. Describe the differences between the basic flip-flop, the SRT flip-flop, and the preferenced flip-flop.
4. Using symbols for the fundamental logic elements, draw schematics for the Boolean expressions below using relay logic, MPL, and nonMPL.
   a. $X+Y+(Z \cdot W) = U$
   b. $XY+X'+Y' = U$
   c. $(X+Y+Z)(XY+X'Z')$
   d. $(X+Y')(Y+Z')(Z+X')(X'+Y')$
   e. $XY'+XZ+XY$
   f. $XYZ+(X+Y)(X+Z)$
   g. $(X+Y)(X+Y')(X'+Z)$
   h. $X'YZ+XY'Z+X'Y'Z+X'YZ'+XY'Z'+X'Y'Z'$
5. Write the Boolean expression for the circuits described by the following schematics:

(a)

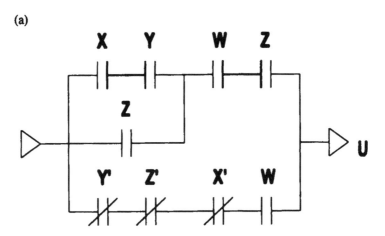

FIGURE 7.35   Circuit A for Problem 5

(b)

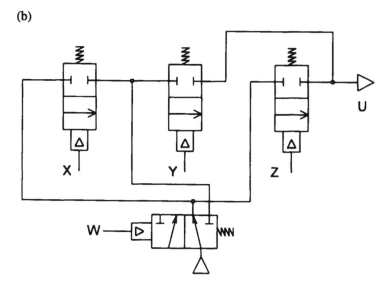

FIGURE 7.36   Circuit B for Problem 5

(c)

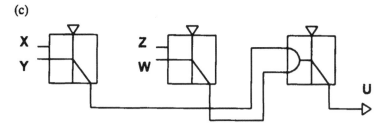

FIGURE 7.37   Circuit C for Problem 5

(d)

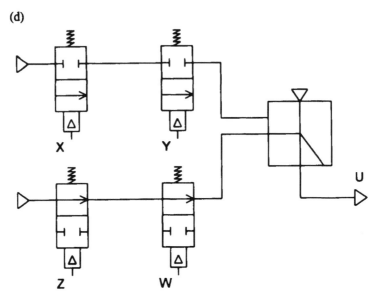

FIGURE 7.38   Circuit D for Problem 5

8 Signal Transmission and Communications

Air logic control rarely functions by itself. It can receive input from external devices, other systems, and humans. It usually has output to external devices, other systems, or humans. Most systems are integrated into other systems, usually electrical controls. Inputs or outputs carry intelligence (data) in the form of pressure signals and may be either analog or digital. Analog signals involve varying pressure to convey intelligence and digital signals merely use constant pressure pulses. The conveyance of these signals and communication with other systems is a critical part of the control system. Transmission of ALC intelligence includes input from sensing devices, output to power devices, and interfacing with other controls.

8.1 DESIGN CONCERNS

Successful and effective signal transmission requires that whatever is received at the output end of a line is the same as what was transmitted at the input end of the line. When conveying ALC intelligence, the mechanism of transmission is a pressure change, usually an increase in pressure at some point in time. This pressure change will be modified during transmission by a variety of factors which include the properties of the pressure signal input and the transmission line conveying the signal. Accuracy of control depends upon the degree of reproduction of the input signal at the point which the signal is to control a device.

Transmission alterations of concern include:

- Attenuation of the signal
- Energy remaining in the signal
- Integrity of pulses
- Time delay of signal arrival at the output

Relationships between transmission parameters and these alterations will be presented for design approximations; however, it is beyond the scope of this book to derive these relationships (Figure 8.1).

8.2 PARAMETERS

Major signal parameters of interest which affect the transmission of pressure signals include (Figure 8.2):

- Pressure amplitude
- Pulse duration
- Pulse frequency
- Distance transmitted

TABLE 8.1
Definitions of Symbols Used in Equations 8.1–8.16

Symbol	Definition
Aa	actuator area (ins²)
C	a constant
CR	compression ratio (pressure in pipe/atmospheric pressure)
Cv	flow coefficient
Ea	actuator energy (lb-ins)
Eo	operating energy (lb-ins)
Fa	actuator force (lbs)
K	ratio of specific heat at constant volume to specific heat at constant pressure (1.4 for air)
L	pipe length (feet)
P	absolute pressure (psia)
Pa	actuator operating pressure (psi)
P1	upstream temperature (psia)
P2	downstream pressure (psia)
Qs	volumetric flow (ACFM)
Sa	actuator stroke (ins)
T	absolute temperature (° Rankine)
V	volume (feet³)
Vl	volume of line or device (feet³)
W	weight (lbs)
W	weight flow (lbs/second)
WF	work factor
d	inside diameter of pipe or tube (inches)
f	pulse frequency (cps)
l	line length (inches)
td	time delay (seconds)
tp	pulse duration (seconds)
v	velocity (feet/sec)
vs	velocity of sound
ΔP	pressure drop (P₁ – P₂) (psi)
ρ	mass density (pounds/feet³)

Two conditions can exist, static and dynamic. Static pressure in a system obeys Pascal's law, and pressure in the system will be undiminished throughout the system regardless of distance from the pressure source or volume of the system. Static conditions require some time after a pressure change to arrive at steady state or static conditions. This time is a function of the volumetric flow into the line and the volume of the line. The time to reach static conditions in seconds can be determined by:

$$t_d = .03472\frac{V_l \cdot CR}{Q_s} \tag{8.1}$$

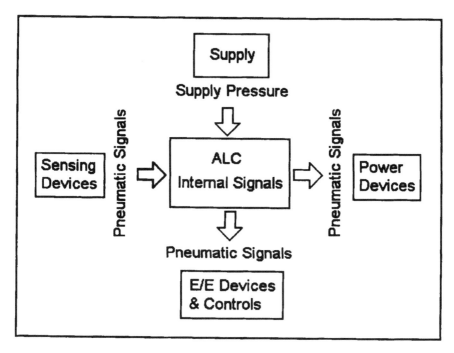

FIGURE 8.1   Flow diagram of ALC transmission functions

Variations in pressure in a compressible fluid produce dynamic conditions. Pascal's law does not apply during dynamic conditions because of the compressibility of air. Three types of dynamic conditions can exist:

- Steady flow
- Pulsing flow
- Transient flow

These conditions result in pressure differentials as a result of losses caused by friction of the conduit wall and compaction (compression) of air molecules. Depending on the type of flow, the resulting pressure differentials will develop different characteristics. Parameters of interest which can be used to analyze the resulting flow characteristics include:

- Pressure attenuation
- Pulse delay
- Phase delay
- Pulse energy

See Table 8.2 Transmission Signal Alterations.

TABLE 8.2
Transmission Signal Alterations

Attenuation
Signal Energy
Pulse Integrity
Time and Phase Delay

8.3 PRESSURE ATTENUATION

Pressure attenuation, or pressure drop, is a function of line properties and flow properties. Line properties which influence attenuation are:

- Cross sectional area
- Length
- Inner surface finish
- Curvature and bends in the line
- Obstructions

Flow properties which effect line losses are:

- Velocity of flow
- Flow type
- Pulse frequency
- Pulse width

Pressure drop in steady state flow lines can be predicted by the famous Harris formula which is repeated from Chapter 2:

$$\Delta P = \frac{c \, L \, Q^2}{CR \, d^5} \qquad \text{(2.23 repeated)}$$

Recall from Chapter 2 that c is function of the conduit in which the flow is occurring and its inside diameter. See Figure 8.2 Pulse parameters.

Pressure drop through devices can be determined by rearranging:

$$Q = 22.48 \cdot C_v \cdot \sqrt{\frac{(P_1 - P_2) \cdot P_2}{T}} \qquad \text{(2.18 repeated)}$$

Which results in:

$$(P_1 - P_2)P_2 = .001979 \; T\left(\frac{Q}{C_v}\right)^2 \qquad \text{(8.2)}$$

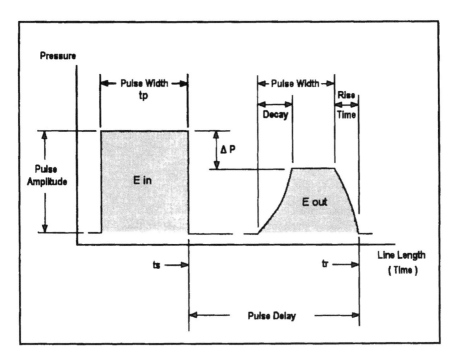

FIGURE 8.2 Pulse parameters

Since P_1 the inlet pressure is usually known, the downstream pressure and the pressure loss can be determined.

8.4 TIME DELAY

Time delay is the time interval between sending a signal pulse at the input to the system and receiving the pulse at the output. Time delay is a measure of the responsiveness of a system to the input signal. Severe time delay can result in incorrect sequencing of controlled devices. Pulse delay can be determined by:

$$t_d = \frac{1}{v} \qquad (8.3)$$

Minimum delay time can be approximated by using the speed of sound as determined by Equation 2.22 for pulse velocity v. In the case of fast rising pulses, the pulse creates a pressure differential, or shock front, as it travels through a conduit so its velocity will exceed sonic velocities. Pulse velocity for fast rising pulses with large pressure differentials (P1/P2) can be determined by:

$$v = v_s \left(.8571 \cdot \frac{\Delta P}{P_2} \right)^{.5} \qquad (8.4)$$

For small pressure fronts with smaller values of P1/P2 the pulse velocity can be determined by:

$$v = v_s\left(1 + .4286 \cdot \frac{\Delta P}{P_2}\right)$$ (8.5)

Note that as the pressure differential decreases, the pulse velocity approaches the speed of sound.

8.5 PHASE DELAY

Phase delay is a measure of a signal's control integrity. Consider a symmetrical control pulse traveling from input to output with a phase delay of 180° then, when the input signal was in the ON state, the output would be in an OFF state. This out of sync condition could frustrate the logic in the control system. Phase delay can be determined by:

$$\theta = 360 f \frac{1}{v}$$ (8.6)

Assuming sonic velocity for v phase delay for small pressure fronts can be approximated by:

$$\theta = .0265 f \, l$$ (8.7)

8.6 PULSE ENERGY

A control pulse may ultimately perform some work on a device such as a valve operator or switching device. To accomplish its intended task, the pulse must contain sufficient energy to do the work required. Energy is a function of force and distance. A control pulse has mass and velocity. From these properties, the energy contained in the pulse can be determined.

$$E = W \, v \, t_p$$ (8.8)

The mass of the pulse can be determined by:

$$W = \frac{V \cdot \rho}{2 \cdot f}$$ (8.9)

And:

$$\rho = 2.7 \frac{P}{T}$$ (8.10)

Remember that P and T are in psia and degrees Rankine.

To determine power delivered equation 11.8 can simply be modified by changing mass of the pulse to mass flow rate as:

$$E = \dot{W} \, v \, t_p \qquad (8.11)$$

And mass flow can be determined by:

$$\dot{W} = 2f \, W = V \, \rho \qquad (8.12)$$

Application of these relationships, quite often, involves operation of power valve actuators or MPL devices. The transmitted pulse must then provide sufficient energy and pressure to operate the valve based on:

$$E_a = F_a \cdot S_a \qquad (8.13)$$

To ensure reliable operation in spite of system variations, additional energy is provided as follows:

$$E_{oper} = E_a \cdot WF = W \cdot v \cdot t_p \qquad (8.14)$$

In addition to required energy to operate the pulse, pressure must be sufficient to develop the necessary operating force:

$$P_a = \frac{F_a}{A_a} \cdot WF = P_a - \Delta P \qquad (8.15)$$

In critical sequencing the pulse flow delay must be considered:

$$t_a = .03472 \frac{S_a \cdot A_a \cdot CR}{Q_s} \qquad (8.16)$$

8.7 SUMMARY

In general, most efficient system operation will result if transmission lines are as short as possible and of sufficient size to provide necessary flow rates with laminar flow. Abrupt directional changes should be avoided to minimize pressure loss, and the number of fittings used should be minimized. Pressure pulses operating valve actuators must be of sufficient amplitude and duration to deliver the required energy for reliable operation.

CHAPTER 8 REVIEW QUESTIONS AND PROBLEMS

1. List the alterations which effect a pressure pulse during transmission.
2. Discuss the effect of major signal parameters on the transmission of a signal pulse.
3. Determine the time to reach steady state pressure after opening a ball valve and flowing air at 5 ACFM in a 100 foot length of 3/4" nominal diameter schedule 40 pipe. Assume the pipe to be at atmospheric pressure and closed at the opposite end.
4. Name 3 types of dynamic conditions in pneumatic transmission lines.
5. Why is pressure attenuation of interest in pneumatic transmission lines?
6. Why is pulse delay of interest in pneumatic transmission lines?
7. Why is phase delay of interest in pneumatic transmission lines?
8. Why is pulse energy of interest in pneumatic transmission lines?
9. Determine the pressure drop for the line in problem 3 with 5 ACFM steady state flow.
10. Determine the pressure drop for the line in problem 3 with 5 SCFM steady state flow.
11. Determine the time delay of a pulse of 20 psig traveling through 50 feet of a $\frac{1}{2}$" diameter stainless steel tubing with a .062" wall. Assume the line to be at atmospheric pressure initially.
12. Determine the velocity of a pulse traveling through a line at sonic speed. Assume standard room temperature.
13. Determine the phase delay of the pulse in Problem 11 if the pulse frequency is 2 pulses per second (pps).
14. Determine the energy in a single pulse in Problem 13.

9 Electrical/Electronic Devices

Air logic systems usually are combined with electrical/electronic devices to exploit the advantages of both types of systems. An understanding of electrical devices is necessary to effectively analyze or design most control systems. Most air logic devices are functionally analogous with their electrical counterparts. The combination of air logic control with the PLC provides a tremendously powerful control system with minimal cost and optimum programming capability. Because of the wide variety of electrical/electronic devices available, only a few of the devices commonly used with pneumatic systems will be presented in this chapter.

9.1 SWITCHES

Switches provide electrical state changes by means of an external actuation. Common types of actuation include manual, pressure, limit (mechanical), electrical (relays), magnetic, temperature, and solid state (Hall effect). Manually operated switches are available in a variety of styles from push-button to key lock. The type of actuator selected is dependent upon the ergonomics of the operator and the required security of the valve operation (Figure 9.1).

Pressure actuation is a common operator type in ALC systems. A typical pressure switch has a low pressure setting and a high pressure setting. The low pressure setting determines the pressure at which the switch closes and the high pressure setting determines the pressure at which the switch opens. Another type has a low pressure setting and a bandwidth setting. As before, the low pressure setting determines the pressure at which the switch closes and the bandwidth setting determines what pressure change can take place before the switch opens (Table 9.1).

Limit actuation is extremely useful for detecting mechanical position. The mechanical operators are available in many designs but in some cases the designer may have to custom design a special shape or size for a specific application.

Standard contact arrangements for most switches include:

SPST NO: single pole single throw, normally open. This arrangement has a single set of normally open contacts (no current through the switch) until a signal is applied to close the contact and allow current to flow.

SPST NC: single pole single throw, normally closed. This arrangement has a single set of normally closed contacts (current flows through the switch) until a signal is applied to the switch to open the circuit and prevent current flow.

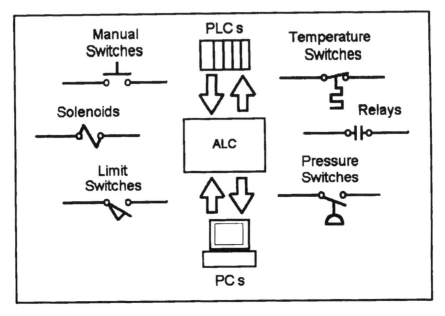

FIGURE 9.1 Electrical/electronic interface with ALC

TABLE 9.1
Air-Electrical Analogies

Air	Electrical
pressure	voltage
pressure drop	voltage drop
flow	current
flow coefficient	1/resistance
$Q \propto Cv\Delta P$	$I \propto E/R$
valves	switches, relays
tubing, pipe, hose	wire, cable
MPL logic	electromechanical logic
fluidic logic	solid state logic
check valve	rectifiers
gages	meters

SPDT: single pole double throw. This arrangement allows two conditions, normally open and normally closed, from a single input simultaneously. An input signal reverses the condition.

DPDT: double pole double throw. This arrangement provides for two independent circuits, either NO or NC.

Single break contact configuration has one pair of contacts for each pole. All current, including inrush in inductive circuits, flows through each contact. This feature limits the current capability of the switch. Double break contacts have two pairs of contacts which the current must flow during make and break. This configuration is a good design for inductive circuits with high inrush such as motor circuits.

Multiple positions and contacts are available in a variety of configurations and ratings. Most switch manufacturers and suppliers provide detailed listings of the switch models including details of the switch configurations and ratings they can provide.

The proper selection of switches must be based on:

- Current flow through the switch
- Voltage of the circuit
- Nature of the current (inductive or resistive)
- Type of operator required

9.2 SOLENOIDS

Solenoids are simply digital linear motors. Applying power displaces a plunger which can be mechanically attached to a valve mechanism changing the state of the valve. Removing power allows the plunger to return to its original position, returning the valve to its former state. The thrust force developed by a solenoid is dependent upon the number of coils of wire in the solenoid, its length, current through it, and the magnetic properties of the plunger. Solenoids are available with thrust forces as small as a fraction of an ounce and as large as several pounds. Electrical actuation of devices is done by means of a solenoids. Solenoids are used primarily as valve and switch actuators.

9.3 RELAYS

Switches operated by solenoids are referred to as relays. The contact arrangement of a relay is referred to as form. The 1 Form A has one set of SPST NO contacts. A 2 Form A relay has two sets of SPST NO contacts. Form B is SPST NC. Contact arrangements up to Form Z are available. Most electrical handbooks contain tables of Form designations for contact arrangements (Figures 9.2 and 9.3).

Relay application requires consideration of:

- Contact Form needed
- Number of contact sets required
- Contact current capacity
- Load type, inductive or resistive
- Contact voltage rating
- Coil operating voltage
- Enclosure requirements

See Figure 9.4 Relay contact configurations and their equivalent MPL values and Figure 9.5 Common switch configurations and their relay equivalents.

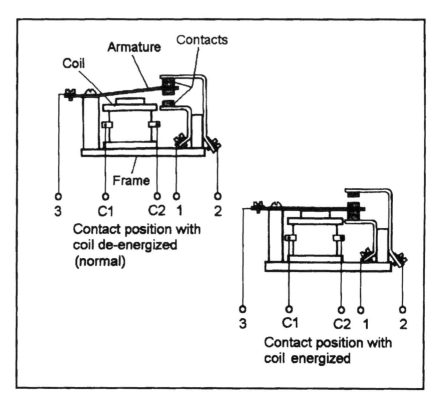

FIGURE 9.2   Line drawing of an electromagnetic relay

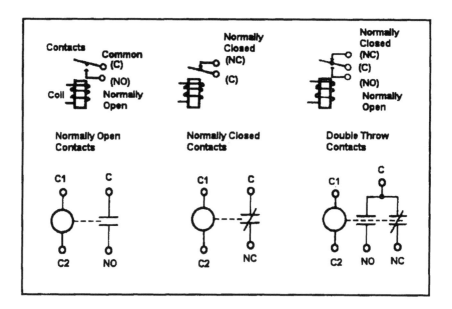

FIGURE 9.3   Relay contact configurations

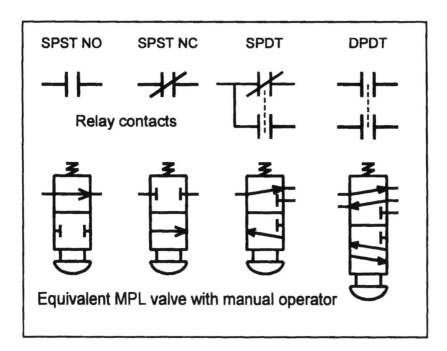

FIGURE 9.4 Relay contact configurations and their equivalent MPL valves

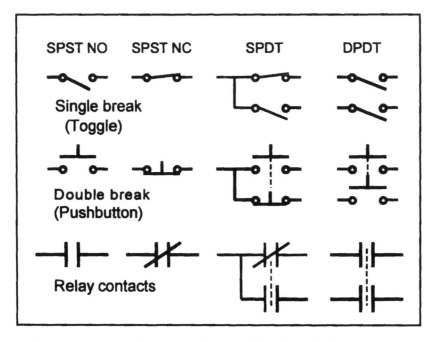

FIGURE 9.5 Common switch configurations and their relay equivalents

9.4 INDICATOR LAMPS

Indicators are used to provide a visual representation of the state of a component or a circuit by illumination or no illumination. Color of the indicator is important for proper interpretation of the indicator state. Generally the following color scheme is suggested:

- Red: danger, abnormal conditions, stop
- Amber: attention, standby
- Green: safe condition, operational, start
- White or clear: normal condition

Three types of indicators are commonly used for industrial control systems:

- Incandescent: this type of indicator is available in several voltages from 6 volts to 110 volts. The lowest voltage bulb possible is the best selection from a safety point of view.
- LED: light emitting diodes are useful as indicators because they can be operated with very low voltages (less than 24 volts) and use very little power. LEDs are available in a variety of colors but are small compared to incandescent bulbs.
- LCD: liquid crystal displays use little power and require very low voltage sources. They do not emit light as the incandescent bulb, and the LED that do so are not as highly visible as the other indicators. Their biggest advantage is they can be made in large panels and produce graphic indications.

9.5 COUNTERS

Counters are memory devices that store input pulses until a preset number is reached and then output a control signal which may terminate or reverse the operation being counted or control another device to sequence it. Electromechanical and solid state counters are available. Since electromechanical devices are prone to fatigue and wear, solid state counters are preferred for most applications. In addition, solid state devices can operate with faster count rates and can be set to count up from zero to the preset value or count down from the preset value to zero. PLCs have counters built-in so they can also provide this function.

9.6 TIMERS

Timers simply measure some preset time before providing an output, usually a switch closure. Two types of timers are available, motor driven and solid state. Motor driven timers can be automatic reset, manual reset, or cycle timers. Automatic reset have a clutch which disengages at the end of the preset cycle time and causes the timing operation to be reset with no external input. The manual reset requires manual reset to zero at the end of each cycle. The repeat cycle timer is a cam operated device which may have multiple switch outputs that can be programmed by orientation and shape of the cams. Solid state timers can be microprocessor based or digital set point. Microprocessor based are the most versatile and have a variety of options for outputs and programmability. They usually have provisions for interfacing with other electronic controls including computers and PLCs. The digital set point timers are simply solid state versions of the motor driven reset timers. Solid state timers are preferred in control system design because they have no parts to wear out and are more reliable than the motor drive types. PLCs also have timing functions built-in so can also provide this function.

9.7   PLCs

Programmable logic controllers (PLCs) have evolved into relatively inexpensive and extremely powerful control devices. The PLC is a microprocessor based control system which is really a dedicated solid state computer. PLCs can handle multiple inputs and provide multiple outputs that can be either analog or digital. A variety of voltages can be used for both input signals and outputs and they are optically isolated to provide ruggedness and reliability. Their versatility is a result of being able to change their operation by easily changed programming techniques. Another major advantage of PLCs is their ability to handle large numbers of inputs and outputs. They can be extremely useful when properly combined with ALC to take advantage of the benefits of both types of controls. Where high density logic is required or interfacing with electrical equipment and devices including networking is necessary, the PLC should be combined with ALC. The designer can easily interface ALC and the PLC since the control logic design and the PLC programming can both be accomplished with ladder logic (Figure 9.6).

9.8   COMPUTERS

Computer control (PC) can be applied as the PLC; however, for most manufacturing applications, computers need to be in hardened enclosures to withstand the shopfloor environment. PLCs are designed to withstand the shopfloor environment. Another disadvantage of computer application to control systems is the unreliability of software, especially the operating systems which are available.

9.9   MOTORS

A variety of electric motors are available to the designer to produce rotating mechanical motion (torque). Power systems use two types of motors AC and DC. DC motors were used in the past because of their variable speed capability. With the development of solid state power circuits with variable frequency output capabilities, motor controls are available which allow variable speed drives to be constructed with AC machines. In many applications, AC variable speed drives have advantages over DC drives. Motion control systems and smaller power motors include servos and steppers. Servo motors are analog in nature and are usually controlled as continuous

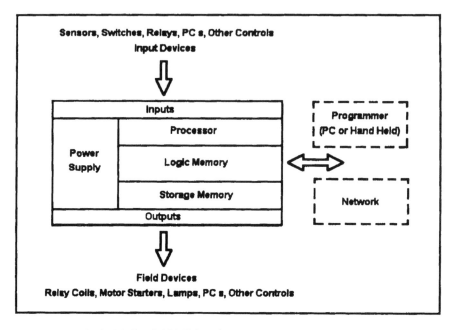

FIGURE 9.6   Typical industrial PLC functions

motion devices. Both AC and DC servos are available. Motion control of the DC servo is simpler than its AC counterpart. Steppers are digital in nature because their motion is generated in discrete steps and can be controlled by providing a measured number of pulses which in turn produce a controlled number of steps. Stepper motors evolved as a result of programmable motion control for machine tools using numerical control (NC). Electric motor application in combination with ALC can best be accomplished by interfacing the ALC with a PLC which in turn can be used as a motor controller.

9.10   SERVO SYSTEMS

Electropneumatic servo systems are used to provide precise and programmable system pressure control. An electropneumatic servo system consists of a pressure sensor, an electronic comparitor/control circuit, and a pressure regulating valve usually in one enclosure. The servo system is a closed loop feedback control which senses the system pressure, then modulates a pressure regulating valve to control the delivered pressure. Applications of this type of control include positional control of pneumatic cylinders and precise speed control of air motors operating at changing input torques.

9.11   TRANSDUCERS

Transducers convert the value of some measured property to a modulated (analog) electrical signal. Commonly properties measured in ALC systems include pressure,

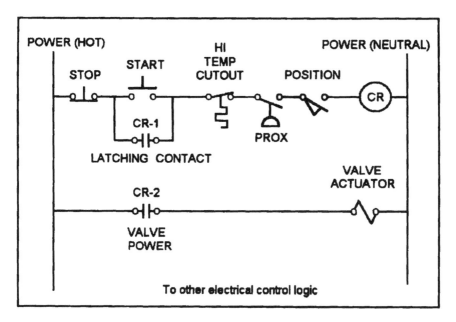

FIGURE 9.7   Basic electrical control circuit

flow rate, temperature, and proximity. It is beyond the scope of this book to discuss the variety of transducers available, but some considerations of common application is necessary. Transducer selection and application requires proper selection of transducer input range, output linearity and repeatability, operational voltage, and environmental constraints. Many transducers are available with integral A/D converters that eliminate many signal transmission problems and allow direct input to computers and PLCs (Figure 9.7).

9.12   BENEFITS OF AIR, ELECTRICAL, HYDRAULIC

Air, hydraulic, and electrical controls have distinct benefits over each other, but they also possess distinct disadvantages. In general the following advantages and disadvantages apply. ALC can be applied with no concern for shock hazard or sparking or arcing. ALC can be used to mimic human control because of its slower speed of control and is usually the least expensive of the three options. Also ALC will not self destruct when exposed to high temperatures as electronic controls are prone to do. The primary disadvantage of ALC is its limited logic density, preventing a large number of components to be used as with electronic devices. ALC is also slow (signals move at nearly the speed of sound) as compared to electronic control (signals move at nearly the speed of light).

The electrical circuit shown in Figure 9.7 illustrates a common arrangement of electrical control. The objective of the scheme is to control the valve actuator when conditions of temperature, proximity, and position exist. The temperature control commonly is a high temperature cut-out to prevent operation when conditions are

TABLE 9.2
Relative Characteristics of Control Systems

Characteristic	Air	Hydraulic	Electrical
Power Source	compressor	pump	generator
Transmission Means	tubing, pipe, hose	tubing, pipe, hose	wire, cable
Transmission Length	line lengths limited to a few feet	long line lengths possible	very long lengths possible
Safety	no sparks, no shock hazard, residual pressure hazard	no sparks, no shock hazard	spark hazard, shock hazard
Cleanliness	clean	messy	clean
Temperature Limits (typical)	250°F max, can be run hot	150°F max, limited by oil temperature	104°F max, easily damaged by excessive heat
Relative Component Size	small somewhat high density logic can be attained	large only low density logic possible	smallest very high logic density possible
Noise	power devices are noisy, may require mufflers	quiet, some flow noise	quiet, usually no noise
System Cost	lowest	highest	usually greater than air
Availability	variety of valves, limited number of fluidic devices available	limited number of devices available	variety of standard devices available

too hot. The proximity condition could be a sensor, such as a back pressure sensor, to detect the presence of a part; in this scheme, a part must be present for actuation to occur. The position limit switch usually senses the extension of a cylinder or other actuator, and in this scheme, the device position must operate the limit switch and hold it closed for the actuator to get power. An additional requirement is for manual operation of the start switch to energize the relay coil CR. Only when all these conditions are satisfied can the valve actuator be powered. The latching contact CR-1 keeps the relay coil powered after the START push-button is released. Powering the relay coil closes the valve power contact CR-2 to apply power to the valve actuator. Other electrical control logic can be included in the same circuit by connecting across the power buses which usually carry 110 volt, 60 hz line power.

Hydraulic controls can produce very high forces which are very positive and are rugged and reliable controls. Also hydraulic systems can be used to produce very precise positioning. Compared to ALC and electronic controls, hydraulic controls are slow, very bulky, and most expensive.

Electrical controls are very versatile and can be constructed with extremely high density logic. Their biggest disadvantages are their relative fragility to the manufacturing environment and their ability to produce sparks and shock hazard (Table 9.2).

CHAPTER 9 REVIEW QUESTIONS AND PROBLEMS

1. What are some advantages of electrical control over ALC?
2. What are some advantages of ALC over electrical controls?
3. What are some disadvantages of electrical control over ALC?
4. What are some disadvantages of ALC over electrical controls?
5. Describe the properties of hydraulic operation.
6. Describe the contact configuration of a SPST switch, a SPDT switch, a DPDT switch.
7. What is the advantage of double break switch contacts?
8. What details are necessary to consider when selecting a relay?
9. List the use for the following color indicator lamps: red, amber, green, white.
10. Discuss the advantage of solid state devices over electromechanical.
11. What is a transducer?
12. Describe the operation and the application of the PLC in ALC systems.

10 Symbology, Schematics, and Flow Diagrams

A working knowledge of graphical symbols for fluid power and electrical devices is necessary to analyze and understand the function of air logic circuits. Symbolism is a language and can be used to communicate information about component connections, component functions, and flow paths. As any language requires the symbols used must have the same meaning to the reader as was intended by the writer, so standardization is necessary to ensure the usefulness of the language. Standardized symbols have been developed by several professional societies including NFPA, ANSI, and ISO. The use of symbols to illustrate a device is referred to as schematic representation. Schematic representation is necessary to illustrate the function of components in a system as well as show the function of the overall system without showing physical details of the components which would be time consuming and unnecessarily complex. Several techniques of schematic representation have been devised to allow the designer to communicate specific details of the functioning of devices and systems.

10.1 STANDARDS

The following standards are presented as references for industry accepted symbols and drawing convention:

- National Fluid Power Association (home page web address is www.nfpa.com)
  NFPA T2.1.3M *Graphic Symbols*
- American National Standards Institute/American Society for Mechanical Engineers (home page web address is www.asme.org)
  ANSI/ASME Y32.10 *USA Standard Graphic Symbols for Fluid Power Diagrams*
- International Organization for Standardization (home page web address is www.iso.org)
  ISO 1219 *Graphical Symbols for Hydraulic and Pneumatic Equipment and Accessories for Fluid Power Transmission*

See Figure 10.1, drawing of a typical power valve with pneumatic operator.

10.2 SYMBOLS

Types of symbols used in drawing circuit diagrams for fluid power systems as well as air logic systems are pictorial, cutaway, and graphic. Graphic symbols are usually

117

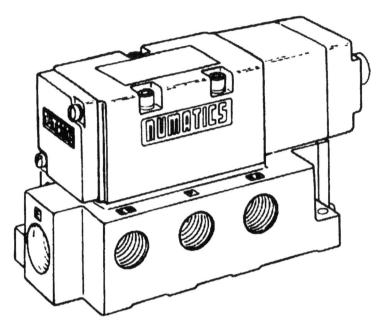

FIGURE 10.1   Power valve with pneumatic operator. (Courtesy of Numatics, Inc., Highland, MI)

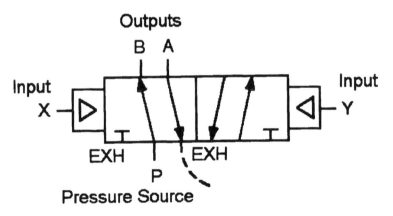

FIGURE 10.2   Graphic symbol of a power valve

the preferred type because they are easiest to draw and emphasize the function of the component. Because of their simplicity relative to other types of symbolism, drawings using them show more clearly the overall function of the described system (Figure 10.2).

Cutaway symbols usually are simplified line drawings of devices that are sectioned to show the internal construction of the device. Most systems are much too complex to describe with cutaway symbols. The benefits of using cutaways usually do not justify the added complexity of the drawing. Cutaway symbols can be useful for unusual devices where understanding the function of the device can be illustrated by the internal details.

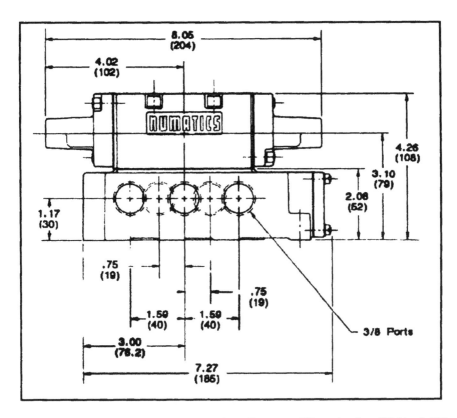

FIGURE 10.3 Line drawing of a power valve. (Courtesy of Numatics, Inc., Highland, MI)

Outline drawings are useful to show the external features of devices. Like cutaways, outline drawings become very complex and cumbersome to work with as the number of devices in a system increases. Outline drawings are useful to identify devices based on their external features and to show the interconnection of devices (Figure 10.3).

10.3 CONVENTION

Symbols should always be drawn to show the state of the device or its location at the start of the working cycle (normal or state with no signal applied). Symbols usually do not show locations of ports, direction of operator travel, or positions of actuators on actual components. Orientation of the symbol in the schematic has no significance to the actual device orientation. Normally, a schematic is drawn from left to right, or top to bottom with operation progressing in the direction drawn. Location of the device symbol in relation to other devices has no significance. The primary function of the schematic is to describe the function of the system (Figure 10.4).

All symbols should conform to an industry accepted standard. Any deviation from standard symbolism needs to be clearly identified and described.

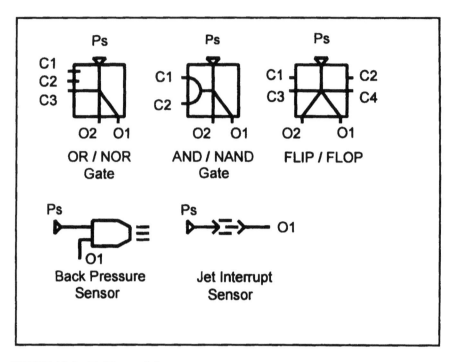

FIGURE 10.4   Fluidic symbols

Only 3 types of lines should be used on any schematic:

Solid line: used for symbol outlines, conductors (tube, hose, or pipe)
Dashed line: used for pilot lines, exhaust, drain lines, and mechanical connections (shaft and rod connectors)
Center or phantom line: used for enclosure outlines and mechanical equipment outlines

See Figure 10.5 Line convention for pneumatic schematics.

Symbols should be simplified to minimize confusion. Any operational detail which is not clearly described by the symbolism should be described by some alternative method.

10.4   SCHEMATICS

Schematic diagrams show groups of devices and how they are connected to each other. A well prepared schematic will also illustrate the function of the circuit and the devices in the circuit.

When preparing a schematic, it is important to keep in mind that the function of the schematic (as well as any engineering drawing) is to communicate. If the devices and their function are not clearly communicated, the schematic is useless.

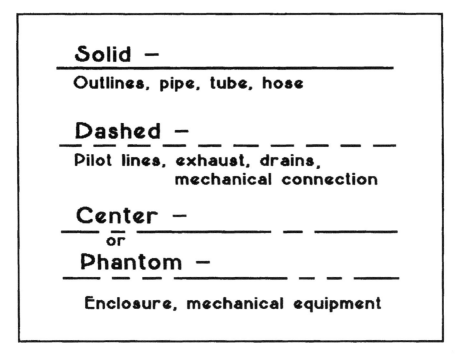

FIGURE 10.5   Line convention for pneumatic schematics

A commonly understood language is necessary to effectively communicate. Use of symbols defined by an industry accepted standard, such as the standards listed at the beginning of this chapter, are recommended to ensure proper communication of circuit information.

Schematics can also function to troubleshoot systems both in the design phase and the maintenance of existing circuits. In the design phase, analysis of the system can save a great deal of time and expense by exposing errors on the schematic. It's a lot easier to correct errors in a schematic than change the real thing after the parts are purchased and assembled.

System maintenance troubleshooting can be expedited using the schematic to determine "what if ..." conditions and even sketching corrections on the drawing. Malfunctioning components can quickly be identified by tracing the operating conditions on the schematic (Figure 10.7 and Figure 10.8).

10.5   ATTACHED METHOD

A single symbol fully describes the function of a device and shows all connections to it. Advantages of the attached method are that a single schematic fully describes the circuit function and shows the interconnection between devices. The disadvantage of this technique is that as complexity of the circuit increases, it becomes difficult to analyze the circuit and determine interaction between devices.

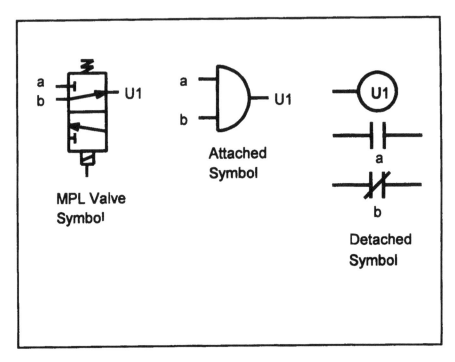

FIGURE 10.6   Attached and detached methods

10.6   DETACHED METHOD

Discrete portions of the devices are split into separate symbols. Detached symbolism
is useful in showing devices with multifunction capabilities and simplifying the representation of complex circuits. The disadvantage of detached symbolism is that several
schematics may be required to describe a single circuit. Also connections between
devices become more difficult to identify and locate (Figure 10.6).

10.7   LOGIC DIAGRAM

The logic diagram uses abstract symbolism which is concerned only with illustrating
the logic of the system. The value of the logic diagram is its simplicity allowing the
reader or designer to see the system logic without concern for physical constraints
(Figures 10.12 and 10.13).

10.8   FLOW CHARTS

The flow chart is a simple way of showing the basic functioning of a control system
without getting lost in details of the circuit. It is recommended that analysis of a
system start with development of a flow chart of the system if one is not available.
Flow charts need not be fancy or formal but should communicate the over all function
of the system with only as much detail as necessary (Figure 10.9).

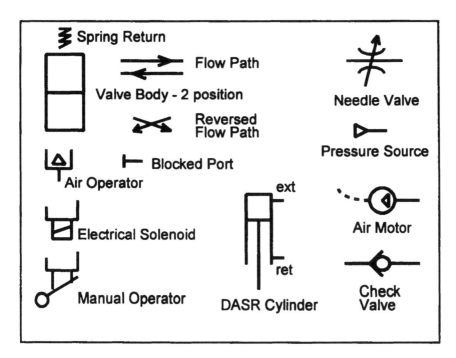

FIGURE 10.7   Basic MPL symbols

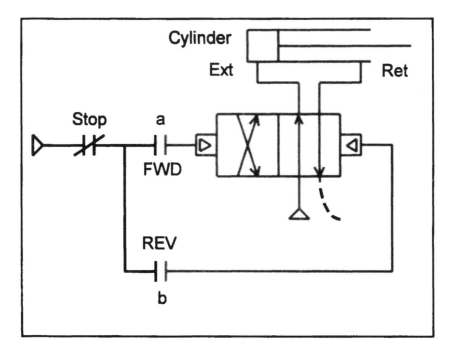

FIGURE 10.8   Circuit schematic using relay logic and MPL

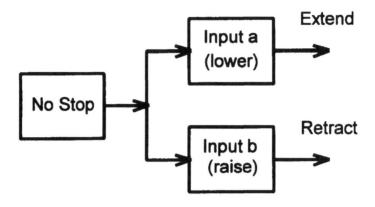

FIGURE 10.9   Typical flow chart

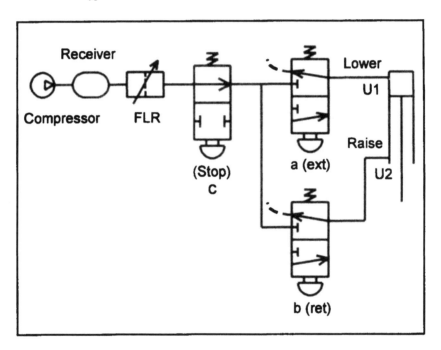

FIGURE 10.10   Schematic of MPL circuit

10.9   CONNECTION DIAGRAMS

Schematics do not necessarily depict the location of connections to a device. To simplify construction of a system connection, diagrams are prepared which function to show the physical arrangement of ports and connections to devices. Use of connection diagrams will also minimize chance of errors in routing and connecting. Labeling of ports on a connection diagram should always correspond to the device manufacturer's labeling and labeling used on the circuit schematic.

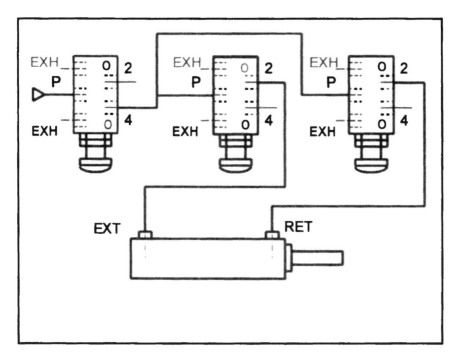

FIGURE 10.11   Connection diagram for MPL circuit

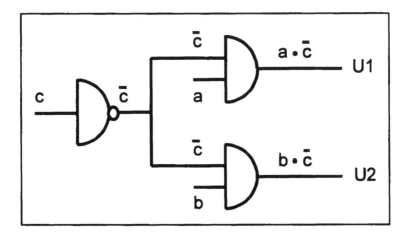

FIGURE 10.12   Logic diagram of an ALC circuit

10.10   MATHEMATICAL MODELS

Even though mathematical models are abstract in nature, their usefulness can be enormous. A great deal of time and aggravation can be saved by modeling a system before putting components together to find they don't perform as expected (Figure 10.14).

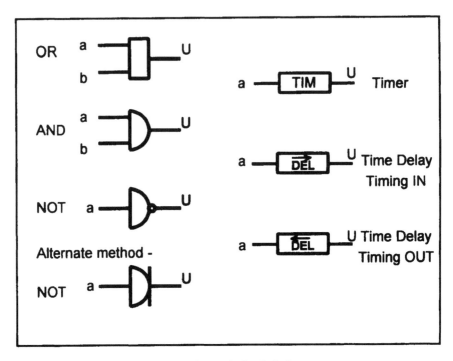

FIGURE 10.13   Logic symbols used for ALC circuit design

$$c' \times a = U1$$
$$c' \times b = U2$$

FIGURE 10.14   Mathematical model of ALC logic

10.11   TRUTH TABLES

Truth tables list, in tabular form, all the input states and show the corresponding output states. The truth table is a convenient way to represent logic statements when there are only a few inputs. When several inputs are involved, the combinations of inputs becomes very large, making the size of the truth table unmanageable (Table 10.1).

10.12   RELAY LOGIC

Relay logic evolved from schematic representation of control systems using electromechanical relays for switching circuits. Since the relay switches contacts on or off, it is a binary device. The basis for ladder diagraming and programming is derived from relay logic symbology. Relay logic symbology can easily be applied to MPL

TABLE 10.1
Typical Truth Table

Inputs			Outputs	
X	Y	Z	U1	U2
0	0	0	0	0
1	0	0	1	0
0	1	0	0	1
1	1	0	1	1
1	0	1	0	0
0	1	1	0	0
1	1	1	0	0

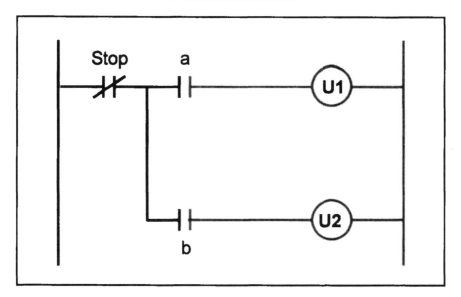

FIGURE 10.15 Ladder diagram for ALC circuit

and used to understand logic circuits developed by combinations of MPL devices. Basically the relay contact is used to represent the state of the valve, and the relay coil is used to represent the output of the circuit. Usually inputs to the MPL device are implied by the specified state of the valve. Relay logic can be used to simplify even very complex systems and can be used to show relative logic between circuit segments as well as discrete devices.

10.13   LADDER DIAGRAMS

Ladder diagrams read from top to bottom and from left to right and may use attached or detached symbols. Usually inputs to the system is shown starting on the left side of the ladder and output from the system are shown on the right side of the rungs (Figure 10.15).

TABLE 10.2
Karnaugh Map

		X+Y			
		0+0	1+0	0+1	1+1
Z	0		U1	U2	*
	1				

*Not useful

10.14 KARNAUGH MAPS

The Karnaugh map is a tabulation of Boolean expressions and allows visualization of the sequences of logic which exist in a logic system. The map is composed of cells or squares. The number of squares is determined by 2 to a power determined by the number of variables, and each square represents a term for all the variables. Two adjacent squares represent terms for one less than the total number of variables. Four adjacent squares represent terms for two less than the total number of variables. Eight adjacent squares represent three less and so on ... Realize that the map can get very complicated as the number of variables increases (Table 10.2).

10.15 SOFTWARE

Computer software is available to assist both the design and analysis of logic controls. Libraries of symbols are available for most CAD programs and very powerful simulation software is available which will show the operation of a circuit as it is drawn (Figure 10.16).

10.16 SUMMARY

Air, electrical, and hydraulic control systems have definite benefits and disadvantages. Air logic control can more nearly mimic human operation because of its inherent response and operating times. ALC is usually simpler in construction than other controls and quite often the least expensive. ALC components are not prone to self destruction as electronic components when overheated and ALC does not possess a shock hazard as electronic systems. The primary disadvantage of ALC is its limited logic density preventing a large number of components to be used. Of course, electronics systems can provide much greater logic densities.

Hydraulic control devices can produce very high forces which are very positive in nature and very reliable compared to electrical and pneumatic systems; however, hydraulic systems are bulky, slow, and expensive compared to both pneumatics and electronics. A tremendous variety of electrical controls is available including packaged systems and controls which are reasonable in cost. Usually the optimum control design is a combination of ALC and electrical devices.

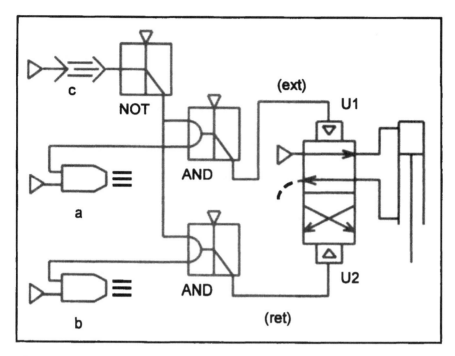

FIGURE 10.16   Fluidic logic schematic

CHAPTER 10 REVIEW QUESTIONS AND PROBLEMS

Problem: Three size boxes of different heights coming down a conveyor line are to be sorted by two back pressure sensors. Three different cylinders are used to eject the boxes to a different que to be filled depending on the outputs of the sensors.

1. Sketch this system using a functional sketch.
2. Sketch a schematic of this system using MPL.
3. Write a mathematical model of this system using Boolean algebra.
4. Show the logic in this system using relay logic.
5. Sketch a flow chart for this system.
6. Sketch a logic diagram for this system.
7. Develop a truth table for this system.
8. Develop a Karnaugh map for this system.

11 ALC System Design

Design can be defined as the process of devising a system, component, or process to satisfy predetermined needs. ALC system design is an integral part of the automated system design for a process. The cardinal rule for designing automated systems is "if you automate a bad process, you will wind up with a bad automated process." Prior to designing the ALC system, the process needs to be analyzed to determine its weaknesses and which areas can be improved before being automated.

Design of an air logic control is primarily a process of selecting and connecting components to form a system which will produce outputs when some combination or sequence of inputs occurs. An air logic control system consists of the entire set of devices and hardware necessary to perform the required control. Design or analysis of air logic systems requires an ability to visualize the system details required to produce the desired end result and some explicit techniques to allow one to understand the effects of the process inputs on system output states.

11.1 THE DESIGN PROCESS

Any design process involves the conversion of ideas to reality. Without direction, this process can be a hit or miss exercise with a great deal of misses. To aid in effectively developing design ideas, the following engineering procedure is recommended:

Define Needs

Determine what is to be accomplished. It is a good idea to document the needs for the design and the details of the project. Any verbal details need to be documented to avoid misunderstanding and embarrassment as the project progresses. Any ambiguity in the needs of the project should be clarified and clearly understood. All variables should be specified with numerical values and documented, especially performance expectations.

Process Analysis

Prior to any automation or control design, the existing process should be studied. Areas of concern include procedures, people, equipment, and environment. Are the procedures currently used appropriate and adequate, are operators knowledgeable and effective, is the equipment reliable and productive, is the environment safe and supportive of the process? Identify any weaknesses in these areas and determine the effect of improvements. Determine the impact of automation on the weaknesses identified.

Approaches to Process Automation

Explore approaches to automating the process and the extent of automation which is practical. Computer Integrated Manufacturing (CIM) is the ultimate level of

TABLE 11.1
Definitions of Symbols Used in Equations 11.1–11.2

Symbol	Definition
ACFM	flow of air through the system (actual cubic feet per minute)
Pavg	average system pressure (psig)
PWR used	power consumption for a specified time in hours (kwatt-hours)
d	demand cycle (hours per hour)
t	time for power consumption (hours)
kW	power usage (kilowatts)
ξ	compressor efficiency factor

automation, but it is not necessarily the optimal approach. CIM involves complete mechanization of a manufacturing process and is theoretically ideal but in many real-world situations is impractical. Increased levels of automation carry an exponentially increasing capital equipment price tag.

Operating Details

List the operating details. Identify inputs and outputs to the process. Flow charts and time lines are valuable tools to present sequences and combinations of process steps and operational requirements.

Control Needs

Determine control needs. Control requirements need to be clearly identified. Any ambiguities need to be clarified before proceeding.

Approaches to Control

Explore approaches to the most effective control system. Choices include electrical/electronic control, ALC, mechanical devices, or hydraulic control. Technology selection should be based on process constraints and requirements compared to the features and benefits of the technology considered. Consider combinations of control technologies for most effective control design.

Logic Development

Model the control system and determine the logic details. Start with the easiest techniques first. The initial logic design usually can be done with relay logic to develop the clearest overall picture of the logic. Several powerful computer simulation packages developed for fluid power applications are available today which can be helpful tools for logic development. Outlining the logic with relay or ladder logic techniques also allows the designer to program the logic into PLC software to test it.

Simplification

Review the developed system and, where possible, combine, eliminate, or redefine the logic. The simplest design will be the most reliable and the least expensive. The goal of simplification should be to reduce the number of process steps, reduce the number of inputs, reduce the number of outputs, reduce the number of components, and reduce the number of connections.

Once the design is acceptable, the mechanics of making the design a reality include:

1. Design documentation
2. Procurement
3. System construction
4. Testing
5. Implementation
6. Follow up

It is beyond the scope of this book to discuss these final steps, but the designer should be aware that the more thoroughly the job is done during the design process, the less trouble and fewer problems will be experienced during the final steps of turning the design into reality.

11.2 PROCESS AUTOMATION

Process automation control systems can involve control of several categories of functions. System complexities develop when several of these functions require integrated control (Figure 11.1).

Moving involves changing the position of a part or a device from point A to point B. The movement may involve linear travel, angular travel, or rotating motion. Development of these types of motion can be accomplished with cylinders for linear motion, rotary actuators for angular movement, and motors for rotation. Control of motion can involve control of starting and stopping the motion, velocity and acceleration of the motion, and distance to be moved.

Holding is required to maintain a part or device in the process during some operation. The critical part of the holding process is the fixture design which must apply the holding forces. Fixture design can reduce the complexity and the criticalness of the control required. Application of the holding force is referred to as clamping. Control parameters include forces required to hold but not deform, orientation of the part or device while being held, sequencing of clamping and unclamping, application of proper force for parts and devices with dimensional changes and differences.

Positioning is the function of putting the part or device at a specified place in space. The location of a part or device during an operation can be critical to the quality of the process results and to the operation of the equipment. Challenges to positioning control are accuracy and precision of location and repeatability of the positioning.

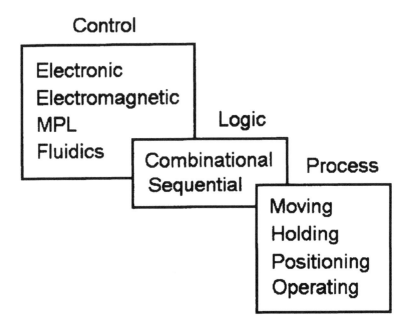

FIGURE 11.1 Process automation control

Operating control usually involves turning the operation on and shutting it off, but may also include control of other parameters such as operating velocity. Sequencing is often a critical parameter in operation control.

11.3 SYSTEM PARAMETERS

The parameters involved in control of automated system functions can include any and every process variable devised, but ALC usually involves the following five parameters:

- Flow
- Pressure
- Sequencing
- Direction
- On/off

Control of these parameters ultimately must result in achieving the specified function of the circuit. The first step in circuit design must always be to define and briefly state the function of the circuit.

11.4 CONTROL TYPES

Open loop control simply involves inputs to the control and outputs from the control. Performance of the control depends on the characteristics of the individual components and devices in the system. Any parameter changes in the system do not affect the control process but may drastically affect the control results (output). Benefits of the open loop control are simplicity and reliability (if control parameters are constant).

Closed loop control involves feedback from the control output to the control input. Feedback signals may be outputs from sensors and transducers which monitor the output variations. Usually the feedback signals or valves require some type of conditioning or modifying. The process of modifying the feedback signal can be complex and critical to the control results. To effectively design feedback control systems, further study in control systems is required.

Combinational control can be open loop or closed loop. Combinational control develops output values for unique sets of multiple input values. Truth tables are useful tools for designing and analyzing Combinational logic.

Sequential control implies operation of a device based on occurrence of another event or events. Sequential control can be open loop or closed loop, but since outputs of this type of control depend on occurrence of other outputs, the system usually is a closed loop type. Sequential control logic always requires some type of memory. Flow charting is a useful tool for initial sequential control design.

Timing control can involve duration of an event or delay of its start or delay of its end.

Combinations of the control types discussed are common. Most automated systems require a number of operations to be controlled to accomplish the intended task so they may utilize a number of controls which are integrated.

11.5   TECHNOLOGY SELECTION

Selection of the proper control technology is an exercise dealing with many choices which have benefits, but ultimately the choice should be based on safety and reliability. Appropriate technical characteristics, of course, should be used for initial screening of choices.

Safety of operation must be the primary consideration for any design. Safety includes safe operation for humans, the system being operated, and other systems and devices. Chapter 15 discusses the safety aspects of control technologies in detail.

Reliability of control in a production environment is critical. Down time for troubleshooting and repair are expensive exercises. The control system may affect a multitude of processes directly and indirectly, so the system must be designed and built to operate correctly and repeatedly in the production environment.

Response time is important to sequencing operations and limits the frequency of operation. Electronic devices are capable of operating with very short response times as low as a few microseconds. Electromechanical, MPL, and fluidic devices are much slower with response times over 1000 times longer. Response times for fluidic devices are typically between 1 and 3 milliseconds and MPL devices usually operate with a response time around 10 milliseconds (Figure 11.2).

Power delivery of both electronic and fluidic devices is relatively low. When providing control to power equipment, secondary devices such as power relays and amplifiers are required. Electromechanical devices can typically deliver over 100 times the power of electronic and fluidic devices. MPL devices can deliver over

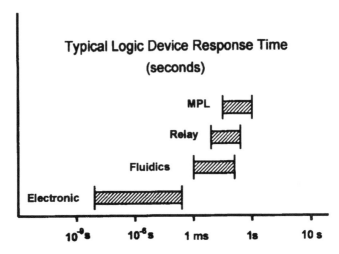

FIGURE 11.2   Response time of logic devices

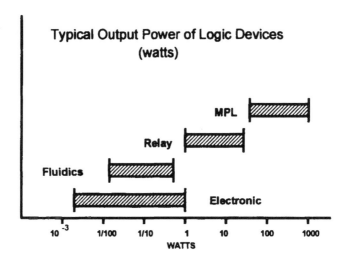

FIGURE 11.3   Logic device power output

1000 times the power of electronic and fluidic devices, so may in many applications directly operate power devices (Figure 11.3).

Power consumption relates to operating cost. In production processes, operating cost can be a significant part of the product cost. Any design should be developed with consideration for minimal power consumption. ALC power consumption is a measure of the air consumption through the system. It is convenient to relate compressed air usage to electrical power in watts. The following equation can be used to approximate the power use for any compressed air system:

$$kW = .003254 \cdot ACFM \cdot P_{avg} \cdot \xi \tag{11.1}$$

Pneumatic systems use power proportional to their cycling rate. When valves are closed, no power is used so power consumption for some time period can be determined by:

$$PWR_{USED} = kW \cdot d \cdot t \qquad (11.2)$$

Solid state electronics devices and fluidic devices use relatively large amounts of power at low cycle rates because they consume a constant amount of power regardless of switching state. At low cycle rates MPL devices have the lowest power consumption. Electromechanical devices also use power only when energized, so are most efficient at low utilization rates.

Compatibility of device technology with other devices and systems should be considered to ensure that system simplicity and reliability are optimized. The proper selection, of course, should be the technology which is most easily interfaced with the power devices and the other controls in the process.

Cost and component price should not be equated. Optimum system cost should be based on, not only the device costs, but also the cost of required associated components and the cost of assembly.

11.6   ALC TECHNOLOGY SELECTION

An important difference between concepts used for air logic devices is the passage size which determines the device operating pressure. High pressure MPL devices have passages typically over .1" diameter and operate at between 30 and 300 psig. Low pressure MPL, which usually are miniaturized devices, have flow passages between .04 and .08" diameter and operate between 3 and 30 psig. Jet deflection fluidic devices have passages between .02 and .04" diameter and operate between .3 and 10 psig. Jet destruction devices have flow passages between .008 and .02 and operate between .2 and 1.8 psig (Figure 11.4).

11.7   SUMMARY

The design process can be summarized by the following steps:

- Define the objective of the system to be designed in one brief statement.
- List the constraints of the system.
- Determine the logic necessary to perform the intended function properly.
- Determine the resources available (money, time, people).
- Design the system.

The cases presented in the following chapters are presented using the procedures discussed in this chapter to provide examples and exercise in their application. Techniques for design include:

- Illustrative schematic
- Algebraic expression of logic

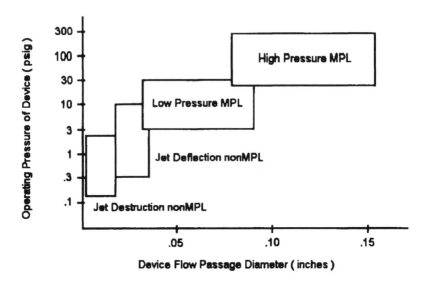

FIGURE 11.4   Operating pressures of ALC devices

TABLE 11.2
Control System Design Tools

Device Schematic
Logic Diagram
Flow Charts
Connection Diagrams
Mathematical Models
Truth Tables
Relay Logic
Ladder Diagrams
Karnaugh Maps
Simulation Software
Time Chart

- Truth table
- Relay logic
- Karnaugh map
- Logic schematic
- Device schematic
- PLC ladder logic

Depending upon the complexity of the system and the preference of the designer, any or all of these techniques may be necessary (See Table 11.2).

CHAPTER 11 REVIEW QUESTIONS AND PROBLEMS

1. Define design.
2. List the major steps in the design process.
3. Why is it important to define the need for a design?
4. What are the major functions for automation control?
5. List 5 process parameters usually controlled by ALC.
6. Define open loop control.
7. Define closed loop control.
8. Define combinational logic.
9. Define sequential logic.
10. List the types of timing control.
11. What are the two major criteria for control technology selection for process control?
12. What is the power consumption for a control system using a flow of 1 ACFM at 30 psig average system pressure with a duty cycle of 75%?
13. What is the normal operating pressure range for high pressure MPL devices?
14. What is the normal operating pressure range for low pressure MPL devices?
15. What is the normal operating pressure range for jet destruction fluidic devices?
16. What is the normal operating pressure range for jet deflection fluidic devices?

12 Applications: Press Control System

Air logic control can be used to solve many control problems. The purpose of this chapter is to put all the information presented in this book together to solve several control problems of increasing complexity. These problems will be presented with alternative design techniques to illustrate their application.

12.1 PRESS CONTROL PROCESS DESCRIPTION

This exercise involves the use of combinational logic with five inputs and one output. The system is a press feeder which pushes a part to be formed under a hydraulic ram. The analysis of this system will be limited to the operation of the feeder only. The operation of the feeder system is as follows:

1. The press ram must be retracted to actuate the feeder
2. No parts or any objects may be under the ram to actuate the feeder
3. The operator protection gate must be down to actuate the feeder
4. Either of two operators may actuate the feeder

See Figure 12.1 Press control operation.

12.2 PROCESS VARIABLES

The required inputs may be identified as:

a. senses position of ram
b. senses presence of a part or object under the ram
c. senses position of operator protection gate
d. operator D input
e. operator E input

The system output is pressure to the valve actuator controlling the feeder cylinder. Output pressure is identified as U. See Figure 12.2 Press control relay logic.

12.3 SYSTEM LOGIC

Three conditions must exist to operate the feeder: the press ram must be retracted, AND no parts may be under the ram, AND the operator protection gate must be down. When these conditions exist, the input from either operator D OR operator E will actuate the feeder.

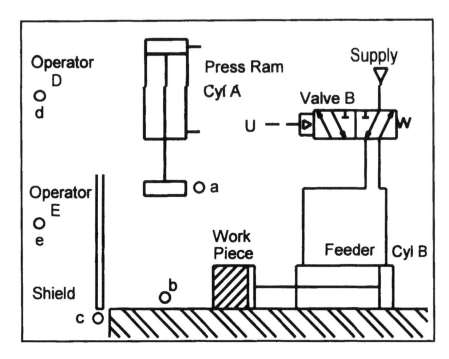

FIGURE 12.1   Press control operation

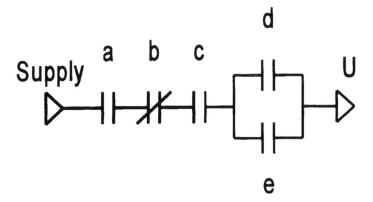

FIGURE 12.2   Press control relay logic

The algebraic expression for this system is:

$$a \bullet b' \bullet c \bullet (d + e) = U \qquad (12.1)$$

Note that sensor b allows flow (passing) unless there is an input to it when a part is present. Input b is required to prevent the feeder from being actuated when a part is already present. b is to be opposite state compared to its input (negated). Therefore b' is chosen to be 1 state with no part present (no input signal), then

TABLE 12.1
Truth Table for Press Feeder Logic

Input States					Output
a	b'	c	d	e	U
0	1	0	0	0	0
1	1	0	0	0	0
1	1	1	0	0	0
1	1	1	1	0	1
1	1	1	0	1	1
1	1	1	1	1	1
1	0	1	1	1	0

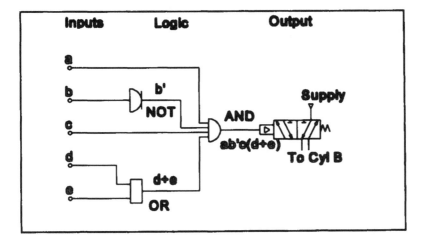

FIGURE 12.3   Logic schematic for press operation

changes to 0 state when a signal is applied (part is present). The control logic can also be illustrated with a truth table showing all the combinations of input and output states (Table 12.1).

Some of the more obvious combinations of inputs have been omitted from the truth table for the sake of brevity. It is recommended that in practice all combinations be tabulated to avoid oversight of any unwanted or inconsistent logic.

The logic may also be illustrated with relay logic as shown in Figure 12.3.

It should be apparent that relay logic can show a very concise picture of the system logic without regard to physical details which can be considered after the necessary logic is developed.

To construct the Karnaugh map for this system, it is necessary to select groups of logic elements to be included for each axis of the map. Logic elements d OR e will be shown on the horizontal axis. a AND b' AND c will be shown on the vertical axis. Realize that each block in this map could be shown by another map illustrating the combinations of the elements in that block (Table 12.2).

TABLE 12.2
Karnaugh Map of Press Feeder Logic

		$a \times b' \times c$							
		010	110	011	111	001	100	101	000
	0+0	0 × 0	0 × 0	0 × 0	1 × 0	0 × 0	0 × 0	0 × 0	0 × 0
		U=0	0	0	0	0	0	0	0
d	0+1	0 × 1	0 × 1	0 × 1	1 × 1	0 × 1	0 × 1	0 × 1	0 × 1
+		0	0	0	1	0	0	0	0
e	1+0	0 × 1	0 × 1	0 × 1	1 × 1	0 × 1	0 × 1	0 × 1	0 × 1
		0	0	0	1	0	0	0	0
	1+1	0 × 1	0 × 1	0 × 1	1 × 1	0 × 1	0 × 1	0 × 1	0 × 1
		0	0	0	1	0	0	0	0

From either the Boolean expression, the truth table, the relay logic, or the Karnaugh map, the logic schematic can be deduced for this system. More complex systems may require all the techniques described to understand the logic. Computer software is also available for logic development. Karnaugh maps may be programmed using most spreadsheet software. A great deal of time and aggravation can be saved when developing complex systems by utilizing programming techniques.

12.4   FLUIDIC CONTROL SYSTEM DESIGN

A very simple control system may be constructed by using fluidic control devices for all logic functions. The fluidic control system would include:

1. A fluidic NOT to negate the b input or use a sensor with a normal 1 state output
2. A 4 input fluidic AND gate
3. A 2 input fluidic OR gate

The output of the AND gate can then be used to operate a pneumatic operator on a DCV to control cylinder B operation.

12.5   MPL CONTROL SYSTEM DESIGN

From the logic schematic, the MPL device schematic may easily be drawn. Note that even though the logic indicates devices d and e following a, b', and c, it is much more practical to put d and e in the beginning of the circuit. This saves piping by allowing only a single line from the operator stations to the press sensors (a, b', and c) (Figure 12.4).

12.6   PLC LADDER LOGIC

PLC ladder logic can easily be developed from relay logic sketches by making a few changes in symbolism and identifying components by addresses as specified by

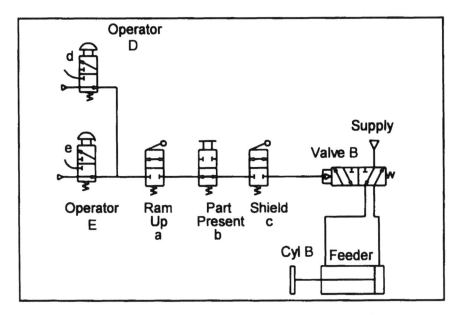

FIGURE 12.4   Press feeder device schematic for MPL control

FIGURE 12.5   Ladder logic for press control

the particular PLC being used. When interfacing this control with a PLC, sensors must provide an electrical contact closure as input to the PLC. The PLC output is a switched voltage to a solenoid operated DCV which controls pressure to the feeder cylinder (Figure 12.5).

CHAPTER 12 REVIEW QUESTIONS AND PROBLEMS

1. Modify the press control logic presented in this chapter to show how the operator controls (inputs e and f) can be changed to a basic two hand safety circuit. Show the new system using:
   a.   Illustrative schematic
   b.   Algebraic expression of logic

    c. Truth table
    d. Relay logic
    e. Karnaugh map
    f. Logic schematic
    g. Device schematic
    h. PLC ladder logic

2. Modify the control to include an additional limit control to limit the travel of the press ram to a predetermined distance from the press table. Show the modified system using:
    a. Illustrative schematic
    b. Algebraic expression of logic
    c. Truth table
    d. Relay logic
    e. Karnaugh map
    f. Logic schematic
    g. Device schematic
    h. PLC ladder logic

3. Where would you locate a safety override such as a key switch to prevent the system from being operated by an unauthorized person? Show the switch using relay logic.

4. Show how you would modify the press control system to reciprocate and operate with a robot removing and replacing the workpiece in the press. Show the new system using:
    a. Illustrative schematic
    b. Algebraic expression of logic
    c. Truth table
    d. Relay logic
    e. Karnaugh map
    f. Logic schematic
    g. Device schematic
    h. PLC ladder logic

5. State the anticipated benefits from using MPL devices for the press control. What are some disadvantages?

6. State the anticipated benefits from using fluidic devices for the press control. What are some disadvantages?

7. State the anticipated benefits from using a PLC for the press control. What are some disadvantages?

8. State the anticipated benefits from using electrical devices for the press control. What are some disadvantages?

13 Applications: Parts Sorting Circuit

The following application provides an example of a simple logic circuit combining combinational and sequential logic. The combinational logic involves the coding of parts by back pressure proximity sensors, and the sequential logic controls the release of parts by an escapement as parts are sensed by back pressure sensors detecting parts at either of three conveyors progressing to an assembly operation.

13.1 PART SORTING PROCESS DESCRIPTION

This control system is presented as a typical product sorting problem which could involve any number of different sizes, shapes, or weights of product, and each product needs to be routed to a different part of the process. For the sake of brevity we will use three different parts: a left hand, a center, and a right hand part. These parts are to be routed to three paths in the process for an assembly operation requiring a left side, a center piece, and a right side. We will assume the parts are being delivered at random by gravity conveyor and that the parts can be delivered to the assembly operation by three separate gravity conveyors. Gravity should always be the first choice for power because it is most reliable and it is free. The entire control operation will be accomplished with ALC operating pneumatic actuators (Figure 13.1).

13.2 PROCESS VARIABLES

The sequence of operation is to be as follows:

1. A part arrives at the part coding sensors (a and b).
2. Cylinder A extends to stop next part and hold it in que.
3. Cylinders B and C are positioned based on code received from part coding sensors.
4. The part is diverted to conveyor 1, 2, or 3.
5. Sensors c, d, or e sense the part as it passes.
6. Cylinder A is retracted to allow the next part to pass.
7. Cycle repeats starting at step 1.

See Figure 13.2 Flow chart for part sorting operation.

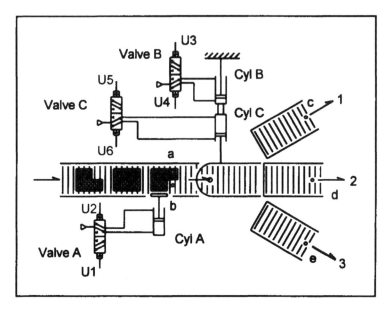

FIGURE 13.1   Part sorting operation

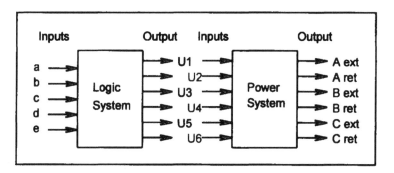

FIGURE 13.2   Flow chart for parts sorting operation

TABLE 13.1
System Part Coding and Path Designation

Part Type	Sensor State a	Sensor State b	Cylinder State B	Cylinder State C	Path
LH	0	1	R	R	1
CENT	1	1	R	E	2
RH	1	0	E	E	3
NONE	0	0	Ret A		Wait

R = Retract   E = Extend

From these operational requirements, the following coding scheme can be developed:

LH	output from sensor a
	no output from sensor b
	take path to conveyor 1
	retract cylinder Band C
CENTER	output from both sensors a and b
	take path to conveyor 2
	cylinder B retracted, cylinder C extended
RH	no output from sensor a
	output from cylinder b
	extend both cylinders B and C
NO PART	no output from sensor a or b
	retract cylinder A

See Table 13.1 System part coding and path designation.

13.3 SYSTEM LOGIC

The following Boolean expressions define the control system:

Retract cylinder A (release a part)—

$$c + d + e = U1$$

Extend cylinder A (hold parts back)—

$$a + b = U2$$

Retract cylinder B—

$$ab + a\bar{b} = a = U4$$

Extend cylinder B—

$$\bar{a}b = U3$$

Extend cylinder C—

$$ab + \bar{a}b = b = U6$$

Retract cylinder C—

$$a\bar{b} = U5$$

Truth tables can be developed for each output individually:

TABLE 13.2
Truth Table for Cylinder A Retraction (U1)

Input States		Output
a	b	U1
0	0	0
1	0	1
0	1	0
1	1	1

TABLE 13.3
Truth Table for Cylinder A Extention (U2)

Input States			Output
c	d	e	U2
0	0	0	0
1	0	0	1
0	1	0	1
0	0	1	1

TABLE 13.4
Truth Table for Cylinder B Extention (U3)

Input States			Output
a	b	b	U3
0	0	1	0
1	0	1	1
0	1	0	0
1	1	0	1

TABLE 13.5
Truth Table for Cylinder B Extention (U4)

Input States		Output
a	b	U4
1	0	0
0	0	0
0	1	0
1	1	1

TABLE 13.6
Truth Table for Cylinder C Retraction (R5)

Input States			Output
a	a	b	U5
0	1	0	0
1	0	0	0
0	1	1	1
1	0	1	1

TABLE 13.7
Truth Table for Cylinder C Extention (U6)

Input States		Output
a	b	U6
0	1	0
0	0	0
1	0	0
1	1	1

The Karnaugh map may be easily modified to show input combinations which result in multiple outputs. Arrows may be added to show sequencing (Table 13.8 and Figure 13.4).

13.4 MPL CONTROL SYSTEM DESIGN

In developing the MPL control note that each sensor requires an MPL device. The logic YES needs only a shutoff valve (sensors c, d, and e) to produce the necessary 0 and 1 states. Functions involving a negated output require 6 port two position valves to provide the negated function simultaneously with the normal function (sensors a and b). Where the negated functions are combined with other functions, they must be isolated with check valves. Without the check valves, AND and OR gates could not be constructed using negated functions. MPL produces the simplest ALC control for this operation (Figure 13.5).

13.5 FLUIDIC CONTROL SYSTEM DESIGN

Fluidic control requires a discrete logic device for each logic function. In addition, because of the low power output of fluidic devices amplifiers are required to operate the power valves adding complexity to the system. Benefits to this design are no moving parts in the control to wear or deteriorate (Figure 13.6).

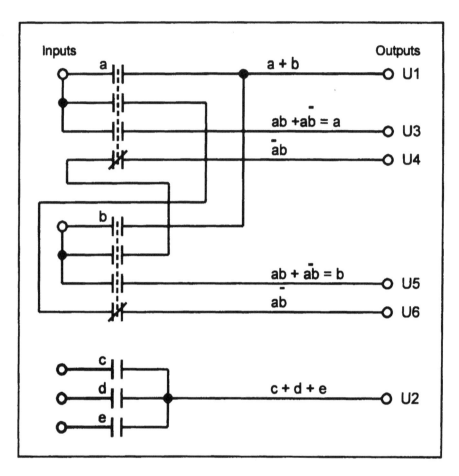

FIGURE 13.3   Relay logic for parts sorting operation

TABLE 13.8
Karnaugh Map for Part Sorting Operation

		c + d + e			
		0+0+0	1+0+0	0+1+0	0+0+1
	00				
ab	01	U2, U4, U5	U1		
	11	U2, U3, U5		U1	
	10	U2, U3, U6			U1

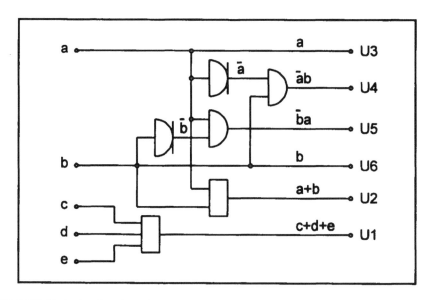

FIGURE 13.4   Logic schematic for part sorting operation

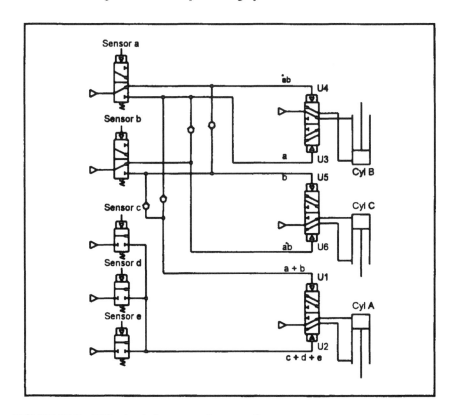

FIGURE 13.5   MPL circuit for part sorting operation

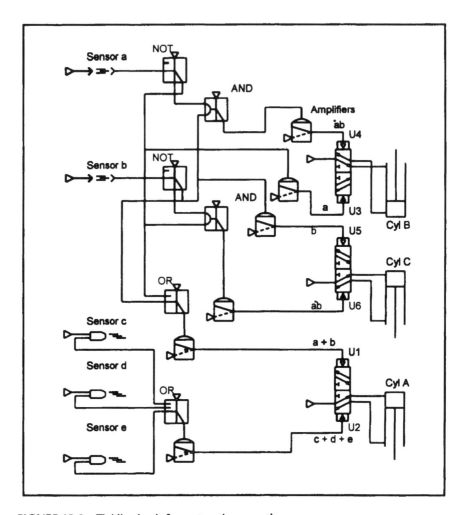

FIGURE 13.6   Fluidic circuit for part sorting operation

13.6   PLC LADDER LOGIC

The PLC control is integrated into one device with electrical outputs which require electrical solenoids to interfere with the operation of pneumatic power valves. This may not be desirable for some applications, especially those in hazardous areas. Another concern is the control must be programmed so it can be easily changed. In a production situation, inadvertent changes can be disastrous, so programmable logic changes may not be desirable (Figure 13.7).

CHAPTER 13 REVIEW QUESTIONS AND PROBLEMS

1. Modify the part sorting logic presented in this chapter to sort only two parts, large boxes and small boxes. Show the new system using:

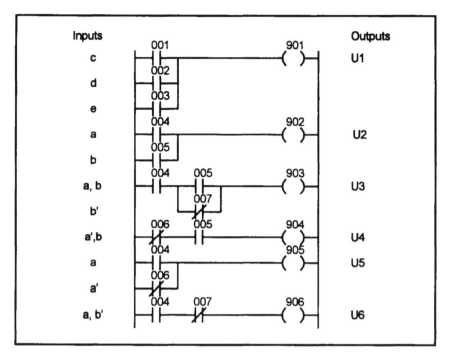

FIGURE 13.7   Ladder logic diagram for part sorting operation

    a. Illustrative schematic
    b. Algebraic expression of logic
    c. Truth table
    d. Relay logic
    e. Karnaugh map
    f. Logic schematic
    g. Device schematic
    h. PLC ladder logic
2. Modify the part sorting logic presented in this chapter to include a color
   sensor (photo cell) to inspect the parts while they are in que. If the color
   value is incorrect Cylinder A will not allow the part to proceed to the
   conveyor. Show the new system using:
    a. Illustrative schematic
    b. Algebraic expression of logic
    c. Truth table
    d. Relay logic
    e. Karnaugh map
    f. Logic schematic
    g. Device schematic
    h. PLC ladder logic

3. How would you modify the part sorting logic presented in this chapter to compensate for parts being placed on the conveyor backwards? Show the new system using:
   a. Illustrative schematic
   b. Algebraic expression of logic
   c. Truth table
   d. Relay logic
   e. Karnaugh map
   f. Logic schematic
   g. Device schematic
   h. PLC ladder logic
4. What control technologies would you use if the parts are very large and heavy?
5. What control technologies would you use if the parts are very small and delicate?
6. State the anticipated benefits from using MPL devices for the parts sorting control. What are some disadvantages?
7. State the anticipated benefits from using fluidic devices for the parts sorting control. What are some disadvantages?
8. State the anticipated benefits from using a PLC for the parts sorting control. What are some disadvantages?
9. State the anticipated benefits from using electrical devices for the parts sorting control. What are some disadvantages?

14 Applications: Bottle Filling System

Air logic control can be applied to most control problems. The purpose of this chapter is to put all the information presented in this book together to solve a control problem involving both combinational and sequential logic. Keep in mind there are alternative approaches to any control problem and usually the most practical design will be a combination of techniques. Even though a combination of electrical control and pneumatics would produce the optimum system, the design presented in this chapter is completely pneumatic to illustrate the possibilities of air logic control.

14.1 BOTTLE FILLING PROCESS DESCRIPTION

The following system provides an example of a sequential logic circuit with a staged process. Combinational logic is used in the system to provide a safety system to prevent spillage and loss of caps as well as staging the fill system. The sequential logic controls the release of bottles by an escapement and movement of the bottles (Figure 14.1).

14.2 PROCESS VARIABLES

We will assume the bottles are delivered to the escapement by a gravity slide and also exit the system by means of a gravity slide. The conveyor is operated by an air motor and all sensors and actuators are pneumatic. Equipment and devices are identified as follows:

A, B	Escapement actuator valves
C	Fluid dispensing valve operator
D	Metering valve operator
E	Capper actuator valve
F	Conveyor motor valve operator
a, b, c, f	Bottle present sensors
d	Fluid level sensor for slowing flow
e	Fluid level sensor for stopping flow

The sequence of operation is as follows:

1. Escapement A releases bottle to isolate.
2. Escapement B releases bottle to the conveyor.

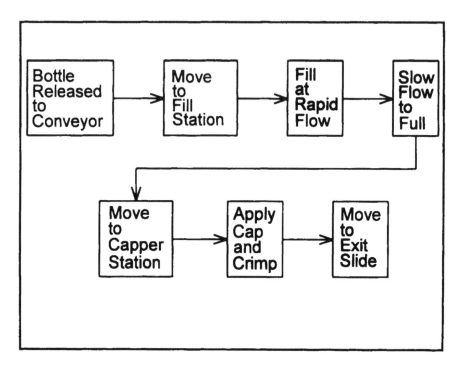

FIGURE 14.1   Bottle fill system flow chart

3. Conveyor F advances to move bottle to the fill station.
4. Fill valve C opens to dispense fluid at fast rate of flow.
5. Sensor d senses fluid level to slow flow.
6. Valve C closes and metering valve D opens to dispense fluid at slow flow.
7. Sensor e detects bottle full to stop flow.
8. Valve D is closed.
9. Conveyor advances to move bottle to capper station and next bottle to fill station. Another bottle is released from the escapement, then another bottle is released to the escapement.

The function of the second stage (slow) flow is to top off the fluid dispensed to provide an accurate fill quantity. Usually second stage is about 1/10 of the fill flow rate (Figure 14.2).

14.3   SYSTEM LOGIC

The following Boolean expressions define the control system:
Move bottle into que by releasing at escapement 1

$$\bar{f} = A \qquad\qquad (14.1)$$

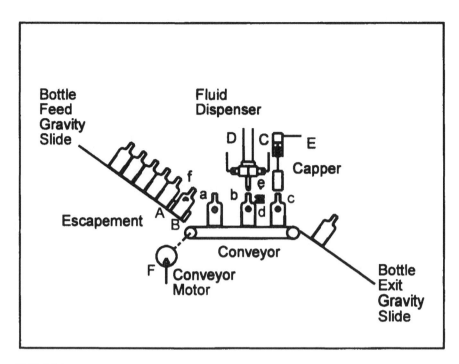

FIGURE 14.2 Bottle filling system

Move bottle onto conveyor by releasing at escapement 2

$$\bar{a} = B \qquad\qquad (14.2)$$

Move bottle to fill station by turning on conveyor motor

$$(\bar{a} + \bar{b} + \bar{c} + e) \cdot d = F \qquad\qquad (14.3)$$

Turn on fast flow to fill bottle

$$b \cdot \bar{d} \cdot \bar{e} = C \qquad\qquad (14.4)$$

Stage flow to fill and stop flow

$$b \cdot d = D \qquad\qquad (14.5)$$

Filled bottle is moved to capper station by control in expression 14.3 and capper places and swages cap

$$c = E \qquad\qquad (14.6)$$

See Table 14.1 Truth Table for Bottle Fill System.

FIGURE 14.1
Truth Table for Bottle Fill System

Inputs (state)										Output & State
a'	b	b'	c	c'	d	d'	e	e'	f '	
0 (1)	0 (0)	0 (1)	0 (0)	0 (1)	0 (0)	0 (1)	0 (0)	0 (1)	0 (1)	A (1) B (1)
									1 (0)	A (0)
1 (0)										B (0)
	1 (1)						1 (0)	1 (0)		C (1)
	1 (1)				1 (1)			1 (0)		D (1)
			1 (1)							E (1)
1 (0)		1 (0)		1 (0)	0 (0)		0 (0)			F (0)
0 (1)		0 (1)		0 (1)	0 (0)			1 (1)		F (0)
0 (1)		0 (1)		0 (1)	1 (1)		0 (0)			F (1)
1 (0)		1 (0)		1 (0)	1 (1)		1 (1)			F (1)

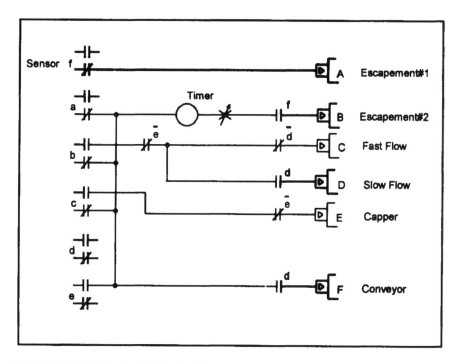

FIGURE 14.3   Relay logic for bottle fill control system

When a control system requires several inputs as this system does, the truth table becomes quite complex and limited in usefulness. Relay logic can be effectively used to illustrate the system logic as shown in Figure 14.3.

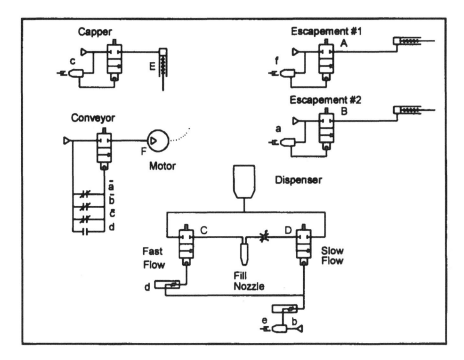

FIGURE 14.4   Schematic of bottle fill control system

14.4   ALC SYSTEM DESIGN

The escapement prevents the weight of successive bottles in line from affecting the motion of the bottle in que. A time delay is needed to ensure the escapement does not open before the bottle in que is completely out of the escapement and on the conveyor before the next bottle is allowed to enter the escapement.

The control system stages the flow allowing precise delivery by shutting off the fluid dispensed at a low flow rate. Dispensing must be done at a rapid rate to meet production rates, but precision of control is sacrificed at high flows. Staging allows most of the fluid to be dispensed at a rapid flow, and then near the end of the cycle, flow is reduced, usually to about 1/10 of the rapid flow to allow shut off at low flow with minimal overflow. Staging is easily accomplished in ALC simply by switching the flow to an alternate line containing a needle vale set to reduce the flow to the desired rate. Staging also reduces hammer which can occur when quickly shutting off flow at high rates.

Staging is also a useful technique to control positioning allowing movement at a rapid rate to near the required position, then slow the velocity usually to 1/10 of the rapid velocity for final slow movement to stop at a precise point. Staging reduces inertial forces due to rapid deceleration from stopping at high velocities (Figure 14.4).

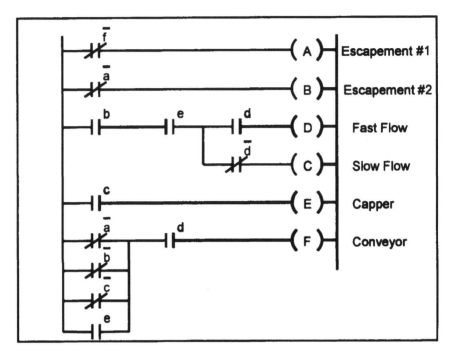

FIGURE 14.5   Ladder logic for bottle fill control system

14.5   PLC LADDER LOGIC

PLC control can take advantage of internal timers to simplify the control hardware. Precise timing is possible by programming the desired value into the controller. This can be a disadvantage because it can easily be changed inadvertently. A definite advantage to combining PLC control with ALC would be to provide for different bottle sizes and fill quantities by simply reprogramming the PLC. Changing the variables of the ALC would require physical changes which may require some adjustment to fine tune each time the process is changed (Figure 14.5).

CHAPTER 14 REVIEW QUESTIONS AND PROBLEMS

1. Modify the bottle filling control logic presented in this chapter to show how the amount of fluid delivered can be controlled by volume of fluid dispensed. Show the new system using:
   a.   Illustrative schematic
   b.   Algebraic expression of logic
   c.   Truth table
   d.   Relay logic
   e.   Karnaugh map
   f.   Logic schematic

    g.   Device schematic

    h.   PLC ladder logic

2. Modify the bottle filling control logic presented in this chapter to show how the amount of fluid delivered can be controlled by weight of fluid dispensed. Show the new system using:

    a.   Illustrative schematic

    b.   Algebraic expression of logic

    c.   Truth table

    d.   Relay logic

    e.   Karnaugh map

    f.   Logic schematic

    g.   Device schematic

    h.   PLC ladder logic

3. Modify the bottle filling control logic presented in this chapter to show how equal amounts of 2 different fluids can be dispensed to fill each bottle. Show the new system using:

    a.   Illustrative schematic

    b.   Algebraic expression of logic

    c.   Truth table

    d.   Relay logic

    e.   Karnaugh map

    f.   Logic schematic

    g.   Device schematic

    h.   PLC ladder logic

4. Modify the bottle filling control logic presented in this chapter using fluidic logic devices wherever possible. Show the new system using:

    a.   Illustrative schematic

    b.   Algebraic expression of logic

    c.   Truth table

    d.   Relay logic

    e.   Karnaugh map

    f.   Logic schematic

    g.   Device schematic

    h.   PLC ladder logic

5. How could the system be modified to eliminate the escapement (A and B)?

6. How could the control system be modified to count the bottles after filling?

7. If the bottles are very tall what would a problem be with the gravity slides? How can this problem be predicted?

8. State the anticipated benefits from using MPL devices for the bottle filling control. What are some disadvantages?

9. State the anticipated benefits from using fluidic devices for the bottle filling control. What are some disadvantages?

10. State the anticipated benefits from using a PLC for the bottle filling control. What are some disadvantages?

11. State the anticipated benefits from using electrical devices for the bottle filling control. What are some disadvantages?

15 Safety

Safety is defined as freedom from danger. Any equipment having an energy source has the potential of producing dangerous situations. Safety must consider both the well-being of humans and the well-being of equipment. Safety is not only a technical problem; it is also an attitude problem. Safety must never be compromised either by design or during operation of any control system. Safety is a serious matter and should be given highest priority. Equipment used in manufacturing, to be useful, often develops large amounts of power and energy. If not properly used, this energy can be harmful and even lethal. To reduce chances of injury, knowledgeable and sensible procedures must be practiced at all times. No list of safety rules can replace knowledge of the equipment and sensible procedures when using it. Workplace safety depends upon knowledge, common sense, and a considerate attitude. Safety involves consideration for yourself, for other workers, and for equipment.

15.1 MAJOR SOURCES OF HAZARDS IN AN INDUSTRIAL ENVIRONMENT

The first step in avoiding and preventing accidents is to identify sources of hazards. The following list only identifies typical sources, so one must be aware that hazards particular to their work environment need to be identified as well.

- Pressure/Energy Sources
  electrical: any voltage over several volts can develop lethal currents and should be considered hazardous.
  mechanical: any moving objects can become pinch points or produce impacts. Any devices which store energy such as springs can be high energy sources.
  hydraulic: pressure in fluids can contain high amounts of energy which can easily produce high amounts of mechanical energy.
  pneumatic: compressed air stores a great deal of energy which can be released and converted into mechanical kinetic energy.
- Chemicals
  Chemical agents used in industrial applications can be toxic, carcinogenic, or extremely corrosive. Some materials (such as mercury) can be absorbed into the body by contact with the skin and cause neurological damage.
- Falling and moving objects
  Gravity always works in one direction and is always working. Heavy objects placed on high shelves or above anything are potential hazards.
- Radiation
  Radiation is an energy source. Any radiation source, either laser light or

gamma radiation, must be handled only by persons with knowledge and
skill to use the source safely.

- Falling, tripping
Tripping and falling can be a result of objects in the way, chemical spills, or
inattentiveness of workers. Good housekeeping can prevent most falls.
- Fire, heat
An ignition source and fuel are required to produce and maintain a fire. Po-
tential ignition sources should be isolated from fuels and fuels should
be maintained properly.
- Steam
Steam is a hazardous energy source because it contains both a great deal
of heat and pressure. Pipes and devices containing steam should be
thermally insulated and identified.
- Noise
Excessive noise, regardless of the source, can damage both the auditory
nerves and neurological cells.
- Rotating equipment
Clothing and body extremities can be entangled with rotating devices if
not properly guarded.
- Fumes, gasses, and airborne particulate matter.
Breathing is a vital bodily function and, if the air contains hazardous ma-
terials, proper breathing apparatus must be used.
- People
People can be safety hazards if they don't possess the proper knowledge
and attitude. No procedures can be effective unless the people involved
have the proper concern and attitude for safety.

15.2   HAZARDS INVOLVING COMPRESSED AIR SYSTEMS

The major hazard involved with compressed air systems is due to the ability of air
to store a great deal of energy (the reason compressed air is useful as an energy
source). The energy stored in compressed air may be inadvertently released by a
mechanical failure of a device such as a rupture in a tube, or a fracture of a fitting,
or other component. Energy released in this fashion may propel pieces of the failed
device at very high velocities (nearly sonic velocities) endangering anyone or anything in close proximity to the device. Because of this possibility, safety glasses
should always be worn when working on or near any pneumatic device. Safety
shields should cover high pressure devices to prevent injuries from failures.

An additional danger is developed by the mechanical devices powered by the
pneumatic system. A person's body parts may be damaged by squashing if caught
between pinch points during system operation. Hands and fingers should never be
placed in pinch points even if the system is not in operation. Unintended startups
are common when testing a system and can result in a great deal of pain and damage
to the careless intruder. If it is necessary to get into the system at a pinch point, the
system should be disabled by removing the main air source or by locking out the

---

TABLE 15.1
Sources of Hazards

Pressure /Energy
Chemicals
Falling or moving objects
Radiation
Falling, tripping
Fire and Heat
Steam
Noise
Fumes and dust
People

---

supply valve. The system should then be vented to atmosphere to relieve any residual pressure (Table 15.1).

15.3 GENERAL SAFETY PRACTICES

Good safety practice is mostly common sense. The notion of common sense is based on good judgement, experience, and knowledge. No set of rules alone can prevent accidents. Documented safety practices are only rules and must be based on common sense and an attitude of concern to be effective. The following general safety practices are presented to provoke a safety attitude. These rules do not cover all situations that must be considered to ensure a safe working environment but are intended to feature major considerations.

1. Always wear safety glasses especially when working on or near pneumatic systems. The compressibility of air, although a beneficial characteristic of pneumatic systems, provides a means of propulsion for parts and pieces of parts when separated from their intended location. Velocity of projectiles formed by escaping air can approach the speed of sound making them dangerous and potentially lethal. These projectiles can easily penetrate the eyeball rendering it inoperable.
2. Never wear neckties, rings, loose fitting clothing, or dangling jewelry around rotating or reciprocating equipment. Never wear rings or metal jewelry when working on electrical equipment.
3. Never resort to horseplay or practical jokes in an industrial environment. Even simple innocent pranks can result in serious injury or even death.
4. Understand the function and operation of equipment before you attempt to operate it. Know what it will do before you push the button.
5. Identify pinch points before you work with equipment. Keep your extremities out of the pinch points.
6. Be familiar with the operating rules for a machine before you operate it. Observe the operating rules; never take chances.

7. Anticipate problems. Know what to do before a problem occurs. Know how to get help.
8. Know where safety equipment is and how to use it. Make sure safety equipment is operational before you have to use it.
9. Know what to shut down and how to shut it down before an emergency.
10. Be sure you are aware of any changes to equipment before you operate it. Check out interlocks and safety devices to ensure they are operational and are performing the function for which they are intended.
11. Respect, but don't fear, equipment. Fear is a result of ignorance, and ignorance is a cause of accidents.
12. Make sure electrical equipment is properly grounded and ground fault interrupters are installed where appropriate. Make sure electrical equipment is fused with appropriate size fuses.
13. When working around live wiring or high voltage circuits, keep one hand in your pocket.
14. It is not sufficient to simply pull the breaker lever and cut the power before working on a system. Always lock the breaker or switch out. Your life may depend on the key to the lock and whether anyone can inadvertently reset the breaker while you are working on it.
15. Fire hazards increase when combustible materials are allowed to accumulate in or near work areas. Keep all combustible materials in suitable storage areas or containers. Be familiar how to react to a fire emergency before one occurs. Know where the best and alternative escape routes are, where extinguishers are located, and where the fire alarm is.
16. Keep work areas neat and clean.
17. Always think, think, think.
18. Ignorance has no place in the industrial control arena or any industrial workplace. Lack of knowledge can result in disastrous situations. People can be badly injured and even killed because of ignorance. The best safeguard against becoming a fatality is to be well-informed.

The workplace must be a safe place. No product or salary is worth hurting someone to get. Being civilized human beings, we must be concerned with our fellow workers' welfare. Safety is everyone's responsibility (Table 15.2).

15.4 OSHA

Occupational Safety and Health Agency (OSHA) is a department of the federal government responsible for setting and enforcing minimum standards for safety in the workplace. These standards are mandated by regulations found in the Code of Federal Regulations 29CFR1910. This code can be found in most libraries or may be purchased from the Government Printing Office. Your tax dollars pay for this effort! When concerned with any aspect of safety, OSHA should be consulted. Many times there is a reluctance to have any contact with OSHA for fear of interference by them, but it has been the author's experience that OSHA should be brought into a situation early before a problem can develop. The attitude should be, "We pay for

TABLE 15.2 **Personal Factors Affecting Safety**
Attitude Knowledge Concern Good Habits

them; we should consult them." Much of this safety discussion is based on information from OSHA documents.

15.5 STATE AGENCIES

Most states have agencies comparable to OSHA but usually have no enforcement power, so they may be consulted without fear of imposition. Most state agencies also have safety training aids and speakers who will do presentations on site. These agencies are paid for by your tax dollars, so their services should be utilized whenever appropriate.

15.6 SUMMARY

No work or activity is more important than human life. It is the responsibility and obligation of the professional designer, technologist, or engineer to work with safety as priority one. Any activity must be pursued with regard for safety as the primary consideration. Errors in judgement must be made in favor of safety and the welfare of humans. Devices and systems must be designed to be fail-safe and be safe when they fail. Compromising safety is unethical, immoral, and illegal.

CHAPTER 15 REVIEW QUESTIONS

1. Explain how you would handle a situation involving a fellow worker deliberately causing unsafe working conditions.
2. Describe the safety considerations involved with a pneumatic power press.
3. Define safety.
4. List the possible energy sources in a manufacturing environment.
5. How can the possibility of being injured by a chemical spill be reduced or prevented?
6. How can the possibility of being injured by a falling object be reduced or prevented?
7. How can the possibility of being injured by a radiation source be reduced or prevented?
8. How can the possibility of being injured by falling or tripping be reduced or prevented?
9. How can the possibility of being injured by a fire be reduced or prevented?

10. How can the possibility of being injured by excessive noise be reduced or prevented?
11. How can the possibility of being injured by rotating equipment be reduced or prevented?
12. How can the possibility of being injured by fumes in the air be reduced or prevented?
13. How can the possibility of being injured by people in the workplace be reduced or prevented?
14. What are some special safety precautions necessary when working around pneumatic equipment and compressed air?

References

Belsterling, Charles A., *Fluidic Systems Design*. New York: John Wiley & Sons, Inc., 1971.

Copeland, Jack, *Artificial Intelligence*. Oxford: Blackwell Publishers, 1993.

Crane Co., *Flow of Fluids—Technical paper no. 410*. Chicago: Crane Co., 1965.

Culbertson, Ralph L., *Air Logic*. Cleveland: Penton Publishing, Inc., 1988.

Ertas Atila, and Jesse C. Jones, *The Engineering Design Process*. New York: John Wiley & Sons, Inc., 1993.

Esposito, Anthony, *Fluid Power with Applications*. 2d ed. Englewood Cliffs, N.J.: Prentice-Hall, Inc., 1988.

Fitch E.C., and J.B. Surjaamadja, *Introduction to Fluid Logic*. Washington: Hemisphere Publishing Corporation, 1978.

Hodge, B.K., and Keith Koenig, *Compressible Fluid Dynamics*. Englewood Cliffs: Prentice-Hall, Inc., 1995.

Hogger, Christopher John, *Introduction to Logic Programming*. Orlando, FL: Academic Press, Inc., 1984.

Parr, Andrew, *Logic Designer's Handbook*. Oxford: Granada Publishing, 1984.

Pease, Dudley A., *Basic Fluid Power*. Englewood Cliffs, New Jersey. Prentice-Hall, Inc., 1987.

Rexford, Kenneth B., *Electric Control for Machines*. Albany: Delmar Publishers, Inc., 1992.

Roth, Jr., Charles H., *Fundamentals of Logic Design*. St. Paul, MN: West Publishing Company, 1985.

Talbott, Edwin M., *Compressed Air Systems*. Lilburn, GA: The Fairmont Press, Inc., 1993.

Index

G

Gas laws 10
Gas limiting velocity 14
Gates 73
Government Printing Office 168
Gravity 147, 165
Gravity slide 157
Grippers 30, 31

H

Hall effect 105
Harris formula 14
Hazardous 165
Hazards 165, 166
Hose 23, 25
Hydraulic
    control 132
    ram 141
Hydraulics 114

I

Ideal gas 10
Impact modulator 65, 66
Indicator lamps 110
Industrial applications 2
Industrial environment 165
Inhibitor 82, 87, 88
Intelligence 97
Intensifiers 32
Interconnection 121
Interface 5, 51, 67
International Organization for
    Standardization 117
Interruptible jet 68, 69
ISO, see International Organization for
    Standardization

J

Jet
    deflection 61–63
    destruction 65, 66
    interaction 61–63

K

Karnaugh maps 128, 138
    applications 144, 151, 152
Kinetic theory 7

L

Ladder diagram 127
Ladder logic 138, 144, 155, 162
Laminar flow 15, 65
Laminar jet 61, 70
Limit actuation 105
Line
    convention 121
    drawing 119
    losses 100
    properties 100
Locking out 166
Logic 35
    circuit 147
    design 132
    development 132
    devices
    diagram 122, 125
    elements 73–93
    functions 1
    schematic 138, 143, 144, 153
    system 128, 158
Lubrication 19
Lubricators 20

M

Manifolding 59
Manifolds 26
Manufacturing systems 3
Mass flow 11, 103
Material selection 27
Mathematical models 125, 126
Mechanical devices 132
Memory 65
Molecular activity 7, 8
Molecules 7
Moles 9
Motion control 111
Motors 133
    air 112
    electric 111
    servo 111
Moving part logic 5, 87, 103, 151
    active 54
    characteristics 135
    circuit 152
    classes 53
    concepts 51–59